ARRL's
TECH
Q&A

Your Quick & Easy Path to your FIRST Ham Radio License!

By Ward Silver, NØAX

Fifth Edition

Contributing Editor:
Mark Wilson, K1RO

Editorial Assistants
Maty Weinberg, KB1EIB

Production Staff:
David Pingree, N1NAS, Senior Technical Illustrator
Jodi Morin, KA1JPA, Assistant Production Supervisor: Layout
Sue Fagan, KB1OKW, Graphic Design Supervisor: Cover Design
Michelle Bloom, WB1ENT, Production Supervisor: Layout

Front Cover Photo: S. Khrystyne Keane, K1SFA

ARRL *The national association for* **AMATEUR RADIO™**
225 Main Street, Newington, CT 06111-1494
www.arrl.org

This book may be used for Technician license exams given beginning July 1, 2010. *QST* and the ARRL Web site (**www.arrl.org**) will have news about any rules changes affecting the Technician license or any of the material in this book.

We strive to produce books without errors. Sometimes mistakes do occur, however. When we become aware of problems in our books (other than obvious typographical errors), we post corrections on the ARRL Web site. If you think you have found an error, please check www.arrl.org for corrections. If you don't find a correction there, please let us know, either using the Feedback Form at the back of this book or by sending e-mail to pubsfdbk@arrl.org.

CONTENTS

Welcome to the diverse group of individuals that make up Amateur Radio! There are nearly 700,000 amateurs, or "hams," in the United States alone and 3,000,000 around the world. Hams come from all walks of life, all ages, and every continent. Hams are busily communicating without regard to the geographic and political barriers that often separate humanity. This is the power of Amateur Radio — to communicate with each other directly, without any other commercial or government systems.

Hams came to Amateur Radio from many walks of life and many interests. Perhaps you intend to provide emergency communications for yourself and your community. Technical experimentation might be your interest or you might be one of the burgeoning "do-it-yourself" community, discovering the pleasures of building, testing, using and learning. Making new friends via the radio, keeping in touch as you travel, or exploring where a wireless signal can take you — these are all valuable and valued parts of the Amateur service.

As you read this book, getting ready to pass your first ham radio licensing exam, be sure to take advantage of the additional material on the ARRL's Web site as suggested in the Introduction. That's the ARRL way of going the extra mile to help you learn about Amateur Radio. "Of, By, and For the Amateur" is the ARRL's motto. By providing this extra information, you will be better prepared to get on the air, have more fun, and be a more effective operator.

Most active radio amateurs in the United States are ARRL members. They realize that since 1914, the ARRL's training, sponsorship of activities, and representation both nationally and internationally are second to none. The book you're reading now, *ARRL's Tech Q&A* is just one of many publications for all levels and interests in Amateur Radio. You don't need a license to join the ARRL — just be interested in Amateur Radio and we are interested in you. It's as simple as that!

David Sumner, K1ZZ
Chief Executive Officer
Newington, Connecticut
March 2010

A Question Pool Committee (QPC) consisting of representatives from the various Volunteer Examiner Coordinators (VECs) prepares the license question pools. The QPC establishes a schedule for revising and implementing new Question Pools. The current Question Pool revision schedule is as follows:

Question Pool	Current Study Guides	Valid Through
Technician (Element 2)	*The ARRL Ham Radio License Manual,* 2nd Edition *ARRL's Tech Q & A,* 5th Edition	**June 30, 2014**
General (Element 3)	*The ARRL General Class License Manual* 6th Edition *ARRL's General Q & A,* 3rd Edition	**June 30, 2011**
Amateur Extra (Element 4)	*The ARRL Extra Class License Manual,* 9th Edition *ARRL's Extra Q & A,* 2nd Edition	**June 30, 2012**

As new Question Pools are released, ARRL will produce new study materials before the effective dates of the new Pools. Until then, the current Question Pools will remain in use, and current ARRL study materials, including this book, will help you prepare for your exams.

As the new Question Pool schedules are confirmed, the information will be published in *QST* and on the ARRLWeb site at **www.arrl.org**.

The Technician License

Earning a Technician Amateur Radio license is a good way to begin enjoying ham radio. Topics covered by the exam provide you with a good introduction to basic radio and there is no difficult math or electronics background required. You are sure to find the operating privileges available to a Technician licensee to be worth the time spent learning about Amateur Radio. After passing the exam, you will be able to operate on every frequency above 50 megahertz that is assigned to the Amateur service. With full operating privileges on those bands, you'll be ready to experience the excitement of Amateur Radio! You also gain privileges on the traditional "shortwave" 80, 40, 15 and 10 meter amateur bands.

Perhaps your interest is in Amateur Radio's long history of providing emergency communications or "emcomm" in time of need. Your experience with computer networks might lead you to explore the many digital modes and networks used in ham radio. If your eyes turn to the stars on a clear night, you might enjoy tracking the amateur satellites and using them to relay your signals to other amateurs around the world! Your whole family can enjoy Amateur Radio, taking part in outdoor activities like ARRL Field Day and mobile operating during a vacation or weekend drive.

An Overview of Amateur Radio

Earning an Amateur Radio license is a special achievement. The nearly 700,000 people in the US who call themselves Amateur Radio operators, or hams, are part of a global fraternity. Radio amateurs provide a voluntary, noncommercial, communication service. This is especially true during natural disasters or other emergencies when the normal lines of communication are out of service. In the aftermath of hurricanes Katrina, Rita, and Wilma in 2005, more than 2000 hams traveled to the stricken areas to establish communications links until normal systems were restored. Thousands more relayed information around the country. In every county and city, organized groups of amateur operators train and prepare to support their communities during disasters and emergencies of every type.

Hams have made many important contributions to the field of electronics and communications, and this tradition continues today. Amateur Radio experimentation is yet another reason many people become part of this self-disciplined group of trained operators, technicians and electronics experts — an asset to any country. Hams pursue their hobby purely for personal enrichment in technical and operating skills, without any type of payment except the personal satisfaction they feel from a job well done!

Radio signals do not know territorial boundaries, so hams have a unique ability to enhance international goodwill. Hams become ambassadors of their country every time they put their stations on the air.

Amateur Radio has been around since the early 1900s. Hams have always been at the forefront of technology. Today, hams relay signals through their own satellites, bounce signals off the moon, relay messages automatically through computerized radio networks and use any number of other "exotic" communications techniques. Amateurs talk from hand-held transceivers through mountaintop repeater stations that can relay their signals to other hams' cars or homes or through the Internet around the world. Hams establish wireless data networks, send their own television signals, and talk with other hams around the world by voice. Keeping alive a distinctive traditional skill, they also tap out messages in Morse code.

The US government, through the Federal Communications Commission (FCC), grants all US Amateur Radio licenses. Amateurs are expected to know more about their equipment and operating techniques because of the tremendous flexibility granted to the Amateur service. Unlike other radio services, amateurs organize their own methods of communication, they are encouraged to build and repair their own equipment, perform experiments with antennas and with radio propagation, and invent their own protocols and networks. The FCC licensing process ensures that amateurs have the necessary operating skill and electronics know-how to use that flexibility wisely and not interfere with other radio services.

Who Can Be a Ham?

The FCC doesn't care how old you are or whether you're a US citizen. If you pass the examination, the Commission will issue you an amateur license. Any person (except the agent of a foreign government) may take the exam and, if successful, receive an amateur license. It's important to understand that if a citizen of a foreign country receives an amateur license in this manner, he or she is a US Amateur Radio operator. (This should not be confused with a reciprocal permit for alien amateur licensee which allows visitors from certain countries who hold valid amateur licenses in their homelands to operate their own stations in the US without having to take an FCC exam.)

License Structure

Anyone earning a new Amateur Radio license can earn one of three license classes — Technician, General and Amateur Extra. Higher-class licenses have more comprehensive examinations. In return for passing a more difficult exam you earn more frequency privileges (frequency space in the radio spectrum). The vast majority of beginners earn the most basic license, the Technician, before beginning to study for the other licenses.

Table 1 lists the amateur license classes you can earn, along with a brief description of their exam requirements and operating privileges. A Technician license gives you the freedom to develop operating and technical skills through on-the-air experience. These skills will help you upgrade to a higher class of license and obtain additional privileges.

The Technician exam, called Element 2, covers some basic radio fundamentals and knowledge of some of the rules and regulations in Part 97 of the FCC Rules. With a little study you'll soon be ready to pass the Technician exam.

Each step up the Amateur Radio license ladder requires the applicant to have passed the lower exams. So if you want to start out as a General class or even an Amateur Extra class licensee, you must first have passed the Technician written exam. Once you have a valid Amateur Radio license, you have credit for all the

Table 1
Amateur Operator Licenses

Class	Written Exam	Privileges
Technician	Basic theory and regulations (Element 2)	All above 50 MHz and limited HF privileges
General	Basic theory and regulations; General theory and regulations (Elements 2 and 3)	All except those reserved for Advanced and Amateur Extra
Amateur Extra	All lower exam elements, plus Amateur Extra theory (Elements 2, 3 and 4)	All amateur privileges

exam elements of that license when you are ready to upgrade. For example, if you hold a Technician license, you will only have to pass the Element 3 General class written exam for a General class license.

Although there are also other amateur license classes, the FCC is no longer issuing new licenses for them. The Novice license was long considered the beginner's license. Exams for this license were discontinued as of April 15, 2000. The FCC also stopped issuing new Advanced class licenses on that date. They will continue to renew previously issued licenses, however, so you will probably meet some Novice and Advanced class licensees on the air.

As a Technician, you can use a wide range of frequency bands — all amateur bands above 50 MHz, in fact. (See **Table 2** and **Figure 1**.) You'll be able to use point-to-point or repeater communications on VHF, use packet radio and other digital modes and networks, even access orbiting satellites or bounce a signal off meteor trails and the Moon! You can use your operating skills to provide public service through emergency communications and message handling.

Station Call Signs

Many years ago, by international agreement, the nations of the world decided to allocate certain call sign prefixes to each country. This means that if you hear a radio station call sign beginning with W or K, for example, you know the station is licensed by the United States. A call sign beginning with the letter G is licensed by Great Britain, and a call sign beginning with VE is from Canada. (All of the amateur call sign prefixes are listed in a table on the ARRL's Web site, **www.arrl.org**.)

The International Telecommunication Union (ITU) radio regulations outline the basic principles used in forming amateur call signs. According to these regulations, an amateur call sign must be made up of one or two characters (the first one may be a numeral) as a prefix, followed by a numeral, and then a suffix of not more than three letters. The prefixes W, K, N and A are used in the United States. When the letter A is used in a US amateur call sign, it will always be with a two-letter prefix, AA to AL. The continental US is divided into 10 Amateur Radio

Table 2
US Amateur Bands
As of February 2010

160 Meters (1.8 MHz)
Avoid interference to radiolocation operations from 1.900 to 2.000 MHz

E,A,G

1.800   1.900   2.000 MHz

80 Meters (3.5 MHz)

3.500 3.600 3.700   MHz

E
A
G
N,T
(200 W)

3.525 3.600   3.800

60 Meters (5.3 MHz)

USB only
2.8 kHz

E,A,G
(50 Watts)

5330.5 5346.5 5366.5 5371.5   5403.5 kHz

General, Advanced, and Amateur Extra licensees may use the following five channels on a secondary basis with a maximum effective radiated power of 50 W PEP relative to a half wave dipole. Only upper sideband suppressed carrier voice transmissions may be used. The frequencies are 5330.5, 5346.5, 5366.5, 5371.5 and 5403.5 kHz. The occupied bandwidth is limited to 2.8 kHz centered on 5332, 5348, 5368, 5373, and 5405 kHz respectively.

40 Meters (7 MHz)

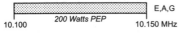

7.000   7.125   7.300 MHz

E
A
G
N,T
(200 W)

7.025   7.125   7.175

Phone and Image modes are permitted between 7.075 and 7.100 MHz for FCC licensed stations in ITU Regions 1 and 3 and by FCC licensed stations in ITU Region 2 West of 130 degrees West longitude or South of 20 degrees North latitude. See Sections 97.305(c) and 97.307(f)(11). Novice and Technician Plus licensees outside ITU Region 2 may use CW only between 7.025 and 7.075 MHz. See Section 97.301(e). These exemptions do not apply to stations in the continental US.

30 Meters (10.1 MHz)
Avoid interference to fixed services outside the US.

E,A,G

200 Watts PEP

10.100   10.150 MHz

20 Meters (14 MHz)

14.000   14.150   14.350 MHz

E
A
G

14.025   14.150 14.225
          14.175

17 Meters (18 MHz)

E,A,G

18.068   18.110   18.168 MHz

15 Meters (21 MHz)

21.000   21.200   21.450 MHz

E
A
G
N,T
(200 W)

21.025   21.200
          21.225
          21.275

12 Meters (24 MHz)

24.890   24.930   24.990 MHz

E,A,G

10 Meters (28 MHz)

28.000 28.300   29.700 MHz

E,A,G
N,T
(200 W)

28.000   28.500

6 Meters (50 MHz)

50.1

E,A,G,T

50.0   54.0 MHz

2 Meters (144 MHz)

144.1

E,A,G,T

144.0   148.0 MHz

1.25 Meters (222 MHz)

E,A,G,T
N (25 Watts)

219.0 220.0   222.0   225.0 MHz

*Geographical and power restrictions may apply to all bands above 420 MHz. See The ARRL Operating Manual for information about your area.

70 cm (420 MHz)*

E,A,G,T

420.0   450.0 MHz

33 cm (902 MHz)*

E,A,G,T

902.0   928.0 MHz

23 cm (1240 MHz)*

1240   1300 MHz

E,A,G,T
N (5 Watts)

1270   1295

ARRL
The national association for
AMATEUR RADIO™

US AMATEUR POWER LIMITS

At all times, transmitter power should be kept down to that necessary to carry out the desired communications. Power is rated in watts PEP output. Except where noted, the maximum power output is *1500 Watts*.

All licensees except Novices are authorized all modes on the following frequencies:

2300-2310 MHz	47.0-47.2	GHz
2390-2450 MHz	76.0-81.0	GHz
3300-3500 MHz	122.25-123.0	GHz
5650-5925 MHz	134-141	GHz
10.0-10.5 GHz	241-250	GHz
24.0-24.25 GHz	All above 275 GHz	

---KEY---

Note:
CW operation is permitted throughout all amateur bands except 60 meters.

MCW is authorized above 50.1 MHz, except for 219-220 MHz.

Test transmissions are authorized above 51 MHz, except for 219-220 MHz

▨ = RTTY and data

■ = phone and image

▨ = CW *only*

≡ = SSB phone

■ = USB phone *only*

▨ = Fixed digital message forwarding systems *only*

E = Amateur Extra
A = Advanced
G = General
T = Technician
N = Novice

See *ARRLWeb* at *www.arrl.org* for more detailed band plans.

ARRL We're At Your Service

ARRL Headquarters
225 Main Street, Newington, CT 0611-1494
www.arrl.org
860-594-0200 (Fax 860-594-0259)
email: hq@arrl.org

Publication Orders
www.arrl.org/catalog
Toll-Free 1-888-277-5289 (860-594-0355)
email: orders@arrl.org

Membership/Circulation Desk
Toll-Free 1-888-277-5289 (860-594-0338)
email: membership@arrl.org

Getting Started in Amateur Radio
Toll-Free 1-800-326-3942 (860-594-0355)
email: Newham@arrl.org

Exams 860-594-0300 vec@arrl.org

QnAbands4 rev. 2/25/2010

call districts (sometimes called areas), numbered 0 through 9. **Figure 2** is a map showing the US call districts.

You may keep the same call sign when you change license class, if you wish. You must indicate that you want to receive a new call sign when you apply for the exam or change your address.

The FCC also has a vanity call sign system. Under this system the FCC will issue an available call sign selected from a list of your preferred call signs. While there is no fee for an Amateur Radio license, there is a fee for the selection of a vanity call sign. The current fee and details of the vanity call sign system are available on the ARRL Web site at **www.arrl.org**.

How to Use This Book

The Element 2 exam consists of 35 questions taken from a pool of more than 350 questions. The *ARRL's Tech Q&A* is designed to help you learn about every question in the Technician exam question pool. Every question is presented just as it is in the question pool and as you will encounter it on the exam. Following every question is a short explanation of the answer.

Each chapter of the book covers one subelement from the question pool, beginning with the FCC Rules and ending with Safety. You may study the questions from beginning to end or select topics in an order that appeals to you.

If you are new to radio, you will probably find it easier to begin with the questions in subelement T3, T4 and T7 to learn about amateur equipment and the basics of radio signals. Then you can move on to the more technical topics covered by T5, T6, T8 and T9. Once you've learned

Figure 1 — This chart details the HF priviliges available to Technician licensees.

about how radios work, the subelements on operating (T2) and FCC Rules (T1) will make more sense. Finish up with T0 — Safety — and you'll be ready for your exam!

The *ARRL Ham Radio License Manual* (*HRLM*) is a good reference companion to the *Tech Q&A*. At the end of the explanation for every question, there is a reference to the page in the *HRLM* where you can find a discussion of the topics associated with the question.

There is additional supplemental material on the ARRL's Web site **www. arrl.org/ham-radio-license-manual** if you need extra help. In particular, there are links to math tutorials and every math problem on the exam is completely worked out to show you how it's done. To make the best use of the online reference material, bookmark the *Ham Radio License Manual* Web site to use as an on-line reference while you study.

The ARRL's New Ham Desk can answer questions emailed to **newham@ arrl.org**. Your question may be answered directly or you might be directed to more instruction material. The New Ham Desk can also help you find a local ham to answer questions. Studying with a friend makes learning the material more fun as you help each other over the rough spots and you'll have someone to celebrate with after passing the exam!

US Call Districts

PR: KP4 VI: KP2

HI and Pacific: **KH6**

AK: **KL7**

ARRLWASmap-g

Figure 2—There are 10 US call sign areas. Hawaii and all Pacific possessions are part of the sixth call area and Alaska is part of the seventh. Puerto Rico and the US Virgin Islands are part of the fourth district.

Earning a License

All US amateur exams are administered by Volunteer Examiners who are certified by a Volunteer-Examiner Coordinator (VEC) that processes the examination paperwork and license applications for the FCC. A Question Pool Committee selected by the Volunteer Examiner Coordinators maintains the question pools for all amateur exams.

Once you make the commitment to study and learn what it takes to pass the exam, you will accomplish your goal. Many people pass the exam on their first try, so if you study the material and are prepared, chances are good that you will soon have your license. It may take you more than one attempt to pass the Technician license exam, but that's okay. There is no limit to how many times you can take it. Many Volunteer Examiner teams have several exam versions available, so you may even be able to try the exam again at the same exam session. Time and available exam versions may limit the number of times you can try the exam at a single exam session. If you don't pass after a couple of tries you will certainly benefit from more study of the question pools before you try again.

License Examinations

The FCC allows Volunteer Examiners to select the questions for an amateur exam, but they must use the questions exactly as they are released by the VEC that coordinates the test session. If you attend a test session coordinated by the ARRL/ VEC, your test will be designed by the ARRL/VEC or by a computer program created by the VEC. The questions and answers will be exactly as they are printed in this book.

Before you can take an FCC exam, you'll have to fill out a copy of the National Conference of Volunteer Examiner Coordinators' (NCVEC) Quick Form 605. This form is used as an application for a new license or an upgraded license. The NCVEC Quick Form 605 is only used at license exam sessions. This form includes some information that the Volunteer Examiner Coordinator's office will need to process your application with the FCC. See **Figure 3**.

You should not use an NCVEC Quick Form 605 to apply for a license renewal or modification with the FCC. Never mail these forms to the FCC, because that will result in a rejection of the application. Likewise, an FCC Form 605 can't be used for a license exam application.

Finding an Exam Session

You can locate upcoming exam sessions in your area by using the ARRL's online Exam Search page. Browse to the ARRL's home page, **www.arrl.org**, and click the "Licensing, Education & Training" button to find complete information about taking a licensing exam. Registration deadlines and the time and location of the exams, are mentioned prominently in publicity releases about upcoming sessions. You can also contact the ARRL/VEC office directly or watch for announcements in the Hamfest Calendar and Coming Conventions columns in *QST*. Many local clubs sponsor exams, so they are another good source of information on exam opportunities.

Taking the Exam

By the time examination day rolls around, you should have already prepared yourself. This means getting your schedule, supplies and mental attitude ready. Plan your schedule so you'll get to the examination site with plenty of time to spare. There's no harm in being early. In fact, you might have time to meet and talk with another applicant which is a great way to calm pretest nerves. Try not to discuss the material that will be on the examination, as this may make you even more nervous. Relax so that you can do your best!

What supplies will you need? First, be sure you bring your current original Amateur Radio license, if you have one. Bring a photocopy of your license, too, as well as the original and a photocopy of any Certificates of Successful Completion of Examination (CSCE) that you plan to use for exam credit. Bring along several

Figure 3—At the test session, the Volunteer Examiners will help you fill out an NCVEC Quick Form 605, which will be filed with the FCC.

sharpened number 2 pencils and two pens (blue or black in). Be sure to have a good eraser. A pocket calculator may also come in handy. You may use a programmable calculator if that is the kind you have, but take it into your exam "empty" (cleared of all programs and constants in memory). Don't program equations ahead of time, because you may be asked to demonstrate that there is nothing in the calculator memory. The examining team has the right to refuse a candidate the use of any calculator that they feel may contain information for the test or could otherwise be used to cheat on the exam.

The Volunteer Examiner team is required to check two forms of identification before you enter the test room. This includes your original Amateur Radio license, if you have one — not a photocopy. A photo ID of some type is best for the second form of ID, but is not required by the FCC. Other acceptable forms of identification include a driver's license, a piece of mail addressed to you or a birth certificate.

The following description of the testing procedure applies to exams coordinated by the ARRL/VEC, although many other VECs use a similar procedure.

Written Test

The examiner will give each applicant a test booklet, an answer sheet and scratch paper. You'll be shown where to sign your name and after that, you're on your own. The first thing to do is read the instructions.

Next, check the examination to see that all pages and questions are there. If not, report this to the examiner immediately. When filling in your answer sheet make sure your answers are marked next to the numbers that correspond to each question.

Go through the entire exam, and answer the easy questions first. Next, go back to the beginning and try the harder questions. Leave the really tough questions for last.

If you don't know the answer to a question, make your best guess. There is no additional penalty for answering incorrectly. If you have to guess, do it intelligently: At first glance, you may find that you can eliminate one or more "distractors." Of the remaining responses, more than one may seem correct; only one is the best answer, however. To the applicant who is fully prepared, incorrect distractors to each question are obvious. Nothing beats preparation!

After you've finished, check the examination thoroughly. You may have read a question wrong or goofed in your arithmetic. Don't be overconfident. There's no rush, so take your time. Think and check your answer sheet. When you feel you've done your best, return the test booklet, answer sheet and scratch pad to the examiner.

The Volunteer Examiner team will grade the exam while you wait. The passing mark is 74%. (That means 26 out of 35 questions correct with up to 9 incorrect answers on the Element 2 exam.) You will receive a Certificate of Successful Completion of Examination (CSCE — see **Figure 4**) showing all exam elements that you pass at that exam session. That certificate is valid for 365 days. Use it as proof that you passed those exam elements so you won't have to take them over again next time you take a license exam.

American Radio Relay League VEC
Certificate of Successful Completion of Examination

ARRL *The national association for* AMATEUR RADIO

NOTE TO VE TEAM:
COMPLETELY CROSS OUT ALL BOXES BELOW THAT DO NOT APPLY TO THIS CANDIDATE.

Test Site (City/State): 02- 11 – 2010   Test Date: Newington, CT

The applicant named herein has presented valid proof for the exam element credit indicated below.

Pre 3/21/87 Technicians
Element 3 credit

CREDIT for ELEMENTS PASSED VALID FOR 365 DAYS
You have passed the written element(s) indicated at right. You will be given credit for the appropriate examination element(s), for up to 365 days from the date shown at the top of this certificate.

EXAM ELEMENTS EARNED

Passed written Element 2
Passed written Element 3
Passed written Element 4

LICENSE UPGRADE NOTICE
If you also hold a valid FCC-issued Amateur radio license grant, this Certificate validates temporary operation with the operating privileges of your new operator class (see Section 97.9[b] of the FCC's Rules) until you are granted the license for your new operator class, or for a period of 365 days from the test date stated above on this certificate, whichever comes first.

NEW LICENSE CLASS EARNED

TECHNICIAN
GENERAL
EXTRA
NONE

LICENSE STATUS INQUIRIES
You can find out if a new license or upgrade has been "granted" by the FCC. For on-line inquiries see the FCC Web at http://wireless.fcc.gov/uls/ ("Click on Search Licenses" button), or see the ARRL Web at http://www.arrl.org/fcc/fcclook.php3; or by calling FCC toll free at 888-225-5322; or by calling the ARRL at 1-860-594-0300 during business hours. Allow 15 days from the test date before calling.

THIS CERTIFICATE IS NOT A LICENSE, PERMIT, OR ANY OTHER KIND OF OPERATING AUTHORITY IN AND OF ITSELF. THE ELEMENT CREDITS AND/OR OPERATING PRIVILEGES THAT MAY BE INDICATED IN THE LICENSE UPGRADE NOTICE ARE VALID FOR 365 DAYS FROM THE TEST DATE. THE HOLDER NAMED HEREON MUST ALSO HAVE BEEN GRANTED AN AMATEUR RADIO LICENSE ISSUED BY THE FCC TO OPERATE ON THE AIR.

Candidate's Signature: *Maria Somma*

Candidate's Name: MARIA SOMMA   Call Sign: KB1KJC (If none, write none)

Address: 225 MAIN ST.

City: NEWINGTON   State CT ZIP 06111

VE #1 *Ennis Harts* N1NAG
Signature   Call Sign

VE #2 *[signature]* KB1DMW
Signature   Call Sign

VE #3 *Perry Green* WY1D
Signature   Call Sign

COPIES: WHITE-Candidate, BLUE-Candidate, YELLOW-VE Team, PINK-ARRL/VEC
MVE 1/2009

Figure 4 — The CSCE (Certificate of Successful Completion of Examination) is your test session receipt that serves as proof that you have completed one or more exam elements. It can be used at other test sessions for 365 days.

Forms and Procedures

To renew or modify a license, you can file a copy of FCC Form 605. In addition, hams who have held a valid license that has expired within the past two years may apply for reinstatement with an FCC Form 605.

Licenses are normally good for 10 years. Your application for a license renewal must be submitted to the FCC no more than 90 days before the license expires. (We recommend you submit the application for renewal between 90 and 60 days before your license expires.) If the FCC receives your renewal application before the license expires, you may continue to operate until your new license arrives, even if it is past the expiration date.

If you forget to apply before your license expires, you may still be able to renew your license without taking another exam. There is a two-year grace period, during which you may apply for renewal of your expired license. Use an FCC Form 605 to apply for reinstatement (and your old call sign). If you apply for reinstatement of your expired license under this two-year grace period, you may not operate your station until your new license is issued.

If you move or change addresses, you should use an FCC Form 605 to notify the FCC of the change. If your license is lost or destroyed, however, just write a letter to the FCC explaining why you are requesting a new copy of your license.

You can ask one of the Volunteer Examiner Coordinators' offices to file your renewal application electronically if you don't want to mail the form to the FCC. You must still mail the form to the VEC, however. The ARRL/VEC office will electronically file application forms. This service is free for any ARRL member.

Electronic Filing

You can also file your license renewal or address modification using the

FCC's Universal Licensing System (ULS) Web site, **www.fcc.gov/uls**. To use ULS, you must have an FCC Registration Number, or FRN. Obtain your FRN by registering with the Commission Registration System, known as CORES.

Described as an agency-wide registration system for anyone filing applications with or making payments to the FCC, CORES will assign a unique 10-digit FCC Registration Number (FRN) to all registrants. All Commission systems that handle financial, authorization of service, and enforcement activities will use the FRN. The FCC says use of the FRN will allow it to more rapidly verify fee payment. Amateurs mailing payments to the FCC — for example as part of a vanity call sign application — would include their FRN on FCC Form 159.

The online filing system and further information about CORES is available by visiting the FCC Web home page, **www.fcc.gov**, and clicking on the Commission Registration System link. Follow the directions on the Web site. It is also possible to register on CORES using a paper Form 160.

When you register with CORES you must supply a Taxpayer Identification Number, or TIN. For individuals, this is usually a Social Security Number. Club stations that do not have an EIN register as exempt.

Anyone can register on CORES and obtain an FRN. You don't need a license to be registered.

Once you have registered on CORES and obtained your FRN, you can proceed to renew or modify your license using the Universal Licensing System by clicking on the "Online Filing" button. Follow the directions provided on the Web page to connect to the FCC's ULS database.

Paper Filing

If you decide to "do the paperwork" on real paper instead of online, you'll need to get a blank FCC Form 605. This is not difficult! You can get FCC Form 605 with detailed instructions by contacting the FCC in any of these ways:

• FCC Forms Distribution Center, 9300 E. Hampton Dr, Capital Heights, MD 20743; tel 800-418-3676 (If you make a written request, write "Form 605" on the envelope.)

• FCC Forms "Fax on Demand" — tel 202-418-0177, ask for form number 000605

• FCC Forms On-Line — **www.fcc.gov/formpage.html** or **ftp.fcc.gov/pub/Forms/Form605**

The ARRL/VEC has created a package that includes the portions of FCC Form 605 that are needed for amateur applications, as well as a condensed set of instructions for completing the form. Write to: ARRL/VEC, Form 605, 225 Main Street, Newington, CT 06111-1494. (Please include a large business-sized stamped self-addressed envelope with your request.)

And Now, Let's Begin

The complete Technician question pool (Element 2) is printed in this book. Each chapter lists all the questions for a particular subelement (such as Operating Procedures — T2). A brief explanation about the correct answer is given after each question.

Table 3 shows the study guide or syllabus for the Element 2 exam as released by the Volunteer-Examiner Coordinators' Question Pool Committee in

February 2010. The syllabus lists the topics to be covered by the Technician exam, and so forms the basic outline for the remainder of this book. Use the syllabus to guide your study.

The question numbers used in the question pool refer to this syllabus. Each question number begins with a syllabus-point number (for example, T0C or T1A). The question numbers end with a two-digit number. For example, question T3B09 is the ninth question about the T3B syllabus topics.

The Question Pool Committee designed the syllabus and question pool so there are the same number of topics in each subelement as there are exam questions from that subelement. For example, three exam questions on the Technician exam must be from the "Operating Procedures" subelement, so there are three groups for that topic. These are numbered T2A, T2B, and T2C. While not a requirement of the FCC Rules, the Question Pool Committee recommends that one question be taken from each group to make the best possible license exams.

Good luck with your studies!

Table 3
Technician Class (Element 2) Syllabus

SUBELEMENT T1 — FCC Rules, descriptions and definitions for the amateur radio service, operator and station license responsibilities
[6 Exam Questions — 6 Groups]

T1A   Amateur Radio services; purpose of the amateur service, amateur-satellite service, operator/primary station license grant, where FCC rules are codified, basis and purpose of FCC rules, meanings of basic terms used in FCC rules

T1B   Authorized frequencies; frequency allocations, ITU regions, emission type, restricted sub-bands, spectrum sharing, transmissions near band edges

T1C   Operator classes and station call signs; operator classes, sequential, special event, and vanity call sign systems, international communications, reciprocal operation, station license licensee, places where the amateur service is regulated by the FCC, name and address on ULS, license term, renewal, grace period

T1D   Authorized and prohibited transmissions

T1E   Control operator and control types; control operator required, eligibility, designation of control operator, privileges and duties, control point, local, automatic and remote control, location of control operator

T1F   Station identification and operation standards; special operations for repeaters and auxiliary stations, third party communications, club stations, station security, FCC inspection

SUBELEMENT T2 — Operating Procedures
[3 Exam Questions — 3 Groups]

T2A   Station operation; choosing an operating frequency, calling another station, test transmissions, use of minimum power, frequency use, band plans

T2B   VHF/UHF operating practices; SSB phone, FM repeater, simplex, frequency offsets, splits and shifts, CTCSS, DTMF, tone squelch, carrier squelch, phonetics

T2C   Public service; emergency and non-emergency operations, message traffic handling

SUBELEMENT T3 — Radio wave characteristics, radio and electromagnetic properties, propagation modes
[3 Exam Questions — 3 Groups]

T3A   Radio wave characteristics; how a radio signal travels; distinctions of HF, VHF and UHF; fading, multipath; wavelength vs. penetration; antenna orientation

T3B   Radio and electromagnetic wave properties; the electromagnetic spectrum, wavelength vs. frequency, velocity of electromagnetic waves

T3C   Propagation modes; line of sight, sporadic E, meteor, aurora scatter, tropospheric ducting, F layer skip, radio horizon

SUBELEMENT T4 — Amateur radio practices and station setup
[2 Exam Questions - 2 Groups]

T4A   Station setup; microphone, speaker, headphones, filters, power source, connecting a computer, RF grounding

T4B   Operating controls; tuning, use of filters, squelch, AGC, repeater offset, memory channels

SUBELEMENT T5 — Electrical principles, math for electronics, electronic principles, Ohm's Law
[4 Exam Questions - 4 Groups]

T5A   Electrical principles; current and voltage, conductors and insulators, alternating and direct current

T5B   Math for electronics; decibels, electronic units and the metric system

T5C   Electronic principles; capacitance, inductance, current flow in circuits, alternating current, definition of RF, power calculations

T5D   Ohm's Law

SUBELEMENT T6 — Electrical components, semiconductors, circuit diagrams, component functions
[4 Exam Groups - 4 Questions]

T6A   Electrical components; fixed and variable resistors, capacitors, and inductors; fuses, switches, batteries

T6B   Semiconductors; basic principles of diodes and transistors

T6C   Circuit diagrams; schematic symbols

T6D   Component functions

SUBELEMENT T7 — Station equipment, common transmitter and receiver problems, antenna measurements and troubleshooting, basic repair and testing
[4 Exam Questions - 4 Groups]

T7A   Station radios; receivers, transmitters, transceivers

T7B   Common transmitter and receiver problems; symptoms of overload and overdrive, distortion, interference, over and under modulation, RF feedback, off frequency signals; fading and noise; problems with digital communications interfaces

T7C   Antenna measurements and troubleshooting; measuring SWR, dummy loads, feedline failure modes

T7D   Basic repair and testing; soldering, use of a voltmeter, ammeter, and ohmmeter

SUBELEMENT T8 — Modulation modes, amateur satellite operation, operating activities, non-voice communications
[4 Exam Questions - 4 Groups]

T8A   Modulation modes; bandwidth of various signals

T8B   Amateur satellite operation; Doppler shift, basic orbits, operating protocols

T8C   Operating activities; radio direction finding, radio control, contests, special event stations, basic linking over Internet

T8D   Non-voice communications; image data, digital modes, CW, packet, PSK31

SUBELEMENT T9 — Antennas, feedlines
[2 Exam Groups - 2 Questions]

T9A   Antennas; vertical and horizontal, concept of gain, common portable and mobile antennas, relationships between antenna length and frequency

T9B   Feedlines; types, losses vs. frequency, SWR concepts, matching, weather protection, connectors

SUBELEMENT T0 — AC power circuits, antenna installation, RF hazards
[3 Exam Questions - 3 Groups]

T0A   AC power circuits; hazardous voltages, fuses and circuit breakers, grounding, lightning protection, battery safety, electrical code compliance

T0B   Antenna installation; tower safety, overhead power lines

T0C   RF hazards; radiation exposure, proximity to antennas, recognized safe power levels, exposure to others

FCC Rules

Your Technician exam (Element 2) will consists of 35 questions taken from the Technician question pool as prepared by the Volunteer Examiner Coordinator's Question Pool Committee. A certain number of questions are taken from each of the 10 subelements. There will be 6 questions from the FCC Rules subelement shown in this chapter. The questions are divided into 6 groups T1A through T1F.

The correct answer (A, B, C or D) is given in bold following the question and the possible responses at the beginning of an explanation section. This convention will be used throughout this book. In addition, at the end of each explanation you'll find the page number where this question is discussed in ARRL's *Ham Radio License Manual*, like this: [*Ham Radio License Manual*, page 7-3].

You'll often see a reference to Part 97 of the Federal Communications Commission rules set in brackets, like this: [97.3(a)(4)]. This tells you where to find the exact wording of the Rules as they relate to that question. You'll find the complete Part 97 Rules on the ARRL Web site at **www.arrl.org**.

SUBELEMENT T1 — FCC Rules, descriptions and definitions for the amateur radio service, operator and station license responsibilities [6 Exam Questions — 6 Groups]

T1A   Amateur Radio services; purpose of the amateur service, amateur-satellite service, operator/primary station license grant, where FCC rules are codified, basis and purpose of FCC rules, meanings of basic terms used in FCC rules

T1A01   For whom is the Amateur Radio Service intended?
A. Persons who have messages to broadcast to the public
B. Persons who need communications for the activities of their immediate family members, relatives and friends
C. Persons who need two-way communications for personal reasons
D. Persons who are interested in radio technique solely with a personal aim and without pecuniary interest

D   [97.3(a)(4)] — The FCC definition of the amateur service is, "A radiocommunications service for the purpose of self-training, intercommunication and technical investigation carried out by amateurs, that is, duly authorized persons interested in radio technique solely with a personal aim and without pecuniary interest." Pecuniary means "related to money or payment," including trade or barter, so amateurs are expected to use Amateur Radio only because they have a personal interest in radio and in radio communications. [*Ham Radio License Manual*, page 7-3]

T1A02   What agency regulates and enforces the rules for the Amateur Radio Service in the United States?

A. FEMA
B. The ITU
C. The FCC
D. Homeland Security

C   [97.5] — Part 97 of the Federal Communication Commission's Rules governs the Amateur Radio Service in the United States. It is the FCC that enforces those rules. [*Ham Radio License Manual*, page 7-2]

T1A03   Which part of the FCC rules contains the rules and regulations governing the Amateur Radio Service?

A. Part 73
B. Part 95
C. Part 90
D. Part 97

D   The Amateur Service is defined by and operates according to the rules in Part 97 of the FCC's rules. [*Ham Radio License Manual*, page 7-1]

T1A04   Which of the following meets the FCC definition of harmful interference?

A. Radio transmissions that annoy users of a repeater
B. Unwanted radio transmissions that cause costly harm to radio station apparatus
C. That which seriously degrades, obstructs, or repeatedly interrupts a radio communication service operating in accordance with the Radio Regulations
D. Static from lightning storms

C   [97.3(A)(23)] — A transmission that disturbs other authorized communications is called harmful interference. FCC Rules define harmful interference as, "Interference which endangers the functioning of a radionavigation service or of other safety communication service operating in accordance with the Radio Regulations." Not all interference is harmful. [*Ham Radio License Manual*, page 8-7]

T1A05   What is the FCC Part 97 definition of a space station?

A. Any multi-stage satellite
B. An Earth satellite that carries one of more amateur operators
C. An amateur station located less than 25 km above the Earth's surface
D. An amateur station located more than 50 km above the Earth's surface

D   [97.3(a)(40)] — Space stations are defined as amateur stations located more than 50 km above the Earth's surface, such as amateur satellites and stations on the International Space Station (ISS) or space shuttle. [*Ham Radio License Manual*, page 8-5]

T1A06   What is the FCC Part 97 definition of telecommand?

A. An instruction bulletin issued by the FCC
B. A one-way radio transmission of measurements at a distance from the measuring instrument
C. A one-way transmission to initiate, modify or terminate functions of a device at a distance
D. An instruction from a VEC

C   [97.3(a)(43)] — Telecommand signals are one-way transmissions intended to initiate, modify or terminate functions of the controlled device. For example, the digital signals used to control a model aircraft are telecommand signals. [*Ham Radio License Manual*, page 6-35]

T1A07   What is the FCC Part 97 definition of telemetry?

A. An information bulletin issued by the FCC
B. A one-way transmission to initiate, modify or terminate functions of a device at a distance
C. A one-way transmission of measurements at a distance from the measuring instrument
D. An information bulletin from a VEC

C   [97.3(a)(45)] — Related to telecommand signals (see question T1A06), telemetry signals are one-way transmissions as well, but send back measurements or status information from a measuring instrument or system. For example, a signal carrying the temperature of a repeater transmitter's enclosure is telemetry. [*Ham Radio License Manual*, page 6-35]

T1A08   Which of the following entities recommends transmit/receive channels and other parameters for auxiliary and repeater stations?

A. Frequency Spectrum Manager
B. Frequency Coordinator
C. FCC Regional Field Office
D. International Telecommunications Union

B   [97.3(a)(22)] — Repeaters input and output frequency pairs are fixed and have a common offset in each region. This enables the maximum number of repeaters to use the limited amount of spectrum. To keep order, a committee of volunteers known as a frequency coordinator recommends transmit and receive frequencies. [*Ham Radio License Manual*, page 7-15]

T1A09   Who selects a Frequency Coordinator?

A. The FCC Office of Spectrum Management and Coordination Policy
B. The local chapter of the Office of National Council of Independent Frequency Coordinators
C. Amateur operators in a local or regional area whose stations are eligible to be auxiliary or repeater stations
D. FCC Regional Field Office

C   [97.3(a)(22)] — The frequency coordinator representatives are selected by the local or regional amateurs eligible to be auxiliary or repeater stations. This ensures participation by local amateurs and is a good example of self-regulation. [*Ham Radio License Manual*, page 7-15]

T1A10   What is the FCC Part 97 definition of an amateur station?

A. A station in an Amateur Radio Service consisting of the apparatus necessary for carrying on radio communications
B. A building where Amateur Radio receivers, transmitters, and RF power amplifiers are installed
C. Any radio station operated by a non-professional
D. Any radio station for hobby use

A   [97.3(a)(5)] — The FCC defines an amateur station as, "A station licensed in the amateur service, including the apparatus necessary for carrying on radio communications." A circular-sounding definition, but remember that the definitions are used in specific legal regulations. What the FCC is saying is that a station that conducts radio communications as required by the amateur service rules in Part 97 meets the definition of an amateur station. [*Ham Radio License Manual*, page 7-3]

T1A11   Which of the following stations transmits signals over the air from a remote receive site to a repeater for retransmission?

A. Beacon station
B. Relay station
C. Auxiliary station
D. Message forwarding station

C   [97.3(a)(7)] — To extend their range and to hear signals blocked by obstacles, repeaters often employ remote receivers. The signals from these receivers are then transmitted by an auxiliary station to the repeater's transmitter site for retransmission. [*Ham Radio License Manual*, page 6-18]

T1B Authorized frequencies; frequency allocations, ITU regions, emission type, restricted sub-bands, spectrum sharing, transmissions near band edges

T1B01 What is the ITU?

A. An agency of the United States Department of Telecommunications Management
B. A United Nations agency for information and communication technology issues
C. An independent frequency coordination agency
D. A department of the FCC

B [97.3(a)(28)] — The International Telecommunication Union (ITU) is an international body of the United Nations that has responsibility for organizing the various radio services on a worldwide basis. This includes arranging international telecommunications treaties, as well as administrative responsibilities such as call signs and frequency allocations. [*Ham Radio License Manual*, page 7-16]

T1B02 North American amateur stations are located in which ITU region?

A. Region 1
B. Region 2
C. Region 3
D. Region 4

B North and South America, including Alaska and Hawaii, form Region 2. [*Ham Radio License Manual*, page 7-16]

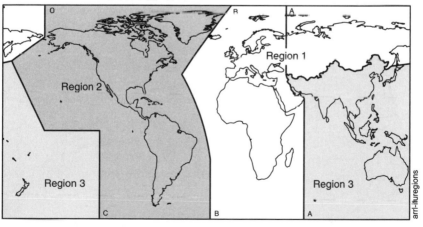

Figure T1-1 — This map shows the world divided into three International Telecommunication Union (ITU) Regions. The US is part of Region 2.

US Call Districts

PR: KP4   VI: KP2

HI and Pacific: **KH6**

AK: **KL7**

ARRLWASmap-g

T1B03   Which frequency is within the 6 meter band?

A. 49.00 MHz
B. 52.525 MHz
C. 28.50 MHz
D. 222.15 MHz

B   [97.301(a)] — The 6 meter band extends from 50.0 to 54.0 MHz in ITU Region 2, which includes North and South America. [*Ham Radio License Manual*, page 7-10]

E = Extra Class
A = Advanced
G = General
P = Technician Plus or Technician who has passed 5 wpm Morse Code Exam
T = Technician

T1B04   Which amateur band are you using when your station is transmitting on 146.52 MHz?

A. 2 meter band
B. 20 meter band
C. 14 meter band
D. 6 meter band

A   [97.301(a)] — The 2 meter band extends from 144.0 to 148.0 MHz in ITU Region 2, which includes North and South America. [*Ham Radio License Manual*, page 7-10]

T1B05 Which 70 cm frequency is authorized to a Technician Class license holder operating in ITU Region 2?

A. 53.350 MHz
B. 146.520 MHz
C. 443.350 MHz
D. 222.520 MHz

C [97.301(a)] — The 70 centimeter band extends from 420.0 to 450.0 MHz in ITU Region 2, which includes North and South America. [*Ham Radio License Manual*, page 7-10]

70 CENTIMETERS **

**Geographical and power restrictions apply to all bands with frequencies above 420 MHz. See Parts 97.301 and 97.303 of the FCC Rules for more infromation.

E,A,G,P,T

420.0   450.0   MHz

= CW, RTTY, data, MCW, test, phone and image   ARRL0554

T1B06 Which 23 cm frequency is authorized to a Technician Class operator license?

A. 2315 MHz
B. 1296 MHz
C. 3390 MHz
D. 146.52 MHz

B [97.301(a)] — The 23 centimeter band extends from 1240 to 1300 MHz in ITU Region 2, which includes North and South America. [*Ham Radio License Manual*, page 7-10]

23 CENTIMETERS **

1270   1295

N
E,A,G,P,T

**Geographical and power restrictions apply to all bands with frequencies above 420 MHz. See Parts 97.301 and 97.303 of the FCC Rules for more information.

1240   1300   MHz
Novices are limited to 5 watts PEP output from 1270 to 1295 MHz.

= CW, RTTY, data, MCW, test, phone and image   ARRL0556

T1B07 **What amateur band are you using if you are transmitting on 223.50 MHz?**

A. 15 meter band
B. 10 meter band
C. 2 meter band
D. 1.25 meter band

D [97.301(a)] — The 1.25-meter band extends from 222 to 225 MHz in ITU Region 2, which includes North and South America. [*Ham Radio License Manual*, page 7-10]

1.25 METERS ***

```
┌──────────────────────────────────────┐
│                                       │ E,A,G,P,T,N
└──────────────────────────────────────┘
222.0                          225.0    MHz
```

Novices are limited to 25 watts PEP output from 222 to 225 MHz.

***219-220 MHz allocated to amateurs on a secondary basis for fixed digital message forwarding systems only and can be operated by all licensees except Novices.

☐ = CW, RTTY, data, MCW, test, phone and image   ARRL0553

T1B08 **What do the FCC rules mean when an amateur frequency band is said to be available on a secondary basis?**

A. Secondary users of a frequency have equal rights to operate
B. Amateurs are only allowed to use the frequency at night
C. Amateurs may not cause harmful interference to primary users
D. Secondary users are not allowed on amateur bands

C [97.303] — A radio service that is designated as the primary service on a band is protected from interference caused by other radio services. A radio service that is designated as the secondary service must not cause harmful interference to, and must accept interference from, stations in a primary service. The amateur service has many different frequency bands. Some of them are allocated on a primary basis and some are secondary. [*Ham Radio License Manual*, page 7-14]

T1B09 **Why should you not set your transmit frequency to be exactly at the edge of an amateur band or sub-band?**

A. To allow for calibration error in the transmitter frequency display
B. So that modulation sidebands do not extend beyond the band edge
C. To allow for transmitter frequency drift
D. All of these choices are correct

D [97.101(a)] — Amateurs are allowed to use any frequency within a band, but you have to be careful when operating near the edge of the band. The rule is that all of your signal must remain within the band. Since your radio displays the carrier frequency, you have to remember to leave room for the signal's sidebands. For example, if your FM voice signal is 15 kHz wide, your carrier frequency (in the center of the signal) should never be less than 7.5 kHz from the band edge. [*Ham Radio License Manual*, page 2-9]

T1B10 Which of the bands available to Technician Class operators have mode-restricted sub-bands?

A. The 6 meter, 2 meter, and 70 cm bands
B. The 2 meter and 13 cm bands
C. The 6 meter, 2 meter, and 1.25 meter bands
D. The 2 meter and 70 cm bands

C [97.305(c)] — Above 50 MHz, the only bands that are divided in any way are 6 meters, 2 meters and 1.25 meters. Both 6 meters and 2 meters have small CW-only sub-bands. 1.25 meters has a separate sub-band from 219-220 MHz in which data signals for fixed message forwarding systems are permitted. Otherwise, there are no mode restrictions in and above the 70 cm band. All of the separation of modes on those bands is according to band plans. [*Ham Radio License Manual*, page 7-13]

T1B11 What emission modes are permitted in the mode-restricted sub-bands at 50.0 to 50.1 MHz and 144.0 to 144.1 MHz?

A. CW only
B. CW and RTTY
C. SSB only
D. CW and SSB

A [97.305 (a)(c)] — A small CW segment is set aside for beacon stations and types of operation that involve very weak signals, such as moonbounce and meteor scatter. [*Ham Radio License Manual*, page 7-13]

T1C **Operator classes and station call signs; operator classes, sequential, special event, and vanity call sign systems, international communications, reciprocal operation, station license and licensee, places where the amateur service is regulated by the FCC, name and address on ULS, license term, renewal, grace period**

T1C01 Which type of call sign has a single letter for both the prefix and suffix?

A. Vanity
B. Sequential
C. Special event
D. In-memoriam

C [97.3(a)(11)(iii)] — Any FCC-licensed amateur or club can also obtain a special 1×1 (1-by-1) call sign such as W3X or K6P for a short-duration special event of significance to the amateur community. These call signs are special because they only have one letter in both the prefix and suffix. [*Ham Radio License Manual*, page 7-21]

T1C02   **Which of the following is a valid US amateur radio station call sign?**

A. KMA3505
B. W3ABC
C. KDKA
D. 11Q1176

B   Every amateur call sign has a prefix and a suffix. In the US, an amateur call sign prefix consists of one or two letters and one numeral. The suffix consists of one to three letters. [*Ham Radio License Manual*, page 7-18]

T1C03   **What types of international communications are permitted by an FCC-licensed amateur station?**

A. Communications incidental to the purposes of the amateur service and remarks of a personal character
B. Communications incidental to conducting business or remarks of a personal nature
C. Only communications incidental to contest exchanges, all other communications are prohibited
D. Any communications that would be permitted on an international broadcast station

A   [97.117] — International treaties limit amateurs to communications for the purposes of the amateur service or remarks of a personal nature. This helps separate commercial and amateur communications. [*Ham Radio License Manual*, page 7-18]

T1C04   **When are you allowed to operate your amateur station in a foreign country?**

A. When the foreign country authorizes it
B. When there is a mutual agreement allowing third party communications
C. When authorization permits amateur communications in a foreign language
D. When you are communicating with non-licensed individuals in another country

A   [97.107] — Amateur radio is not authorized by all countries! If amateur radio is authorized, you must be licensed according to that country's regulations. Your US amateur license may suffice if there is a reciprocal operating agreement between the US and the foreign country, enabling you to operate your radio according to the rules of that country. There are three types of agreement: the International Amateur Radio Permit (IARP), the European Conference of Postal and Telecommunications Administration (CEPT) agreement, and international reciprocal agreements between the foreign country and the US. [*Ham Radio License Manual*, page 7-17]

T1C05 **What must you do if you are operating on the 23 cm band and learn that you are interfering with a radiolocation station outside the United States?**

A. Stop operating or take steps to eliminate the harmful interference
B. Nothing, because this band is allocated exclusively to the amateur service
C. Establish contact with the radiolocation station and ask them to change frequency
D. Change to CW mode, because this would not likely cause interference

A [97.303(h)] — Not all US amateur allocations are allocated to amateurs worldwide. Where there are competing allocations and another service is considered the primary user, the amateur service is considered to be secondary use. (See question T1B08.) [*Ham Radio License Manual*, page 7-15]

T1C06 **From which of the following may an FCC-licensed amateur station transmit, in addition to places where the FCC regulates communications?**

A. From within any country that belongs to the International Telecommunications Union
B. From within any country that is a member of the United Nations
C. From anywhere within ITU Regions 2 and 3
D. From any vessel or craft located in international waters and documented or registered in the United States

D [97.5(a)(2)] — Vessels and crafts documented or registered in the United States are bound by US rules and regulations. In addition, you must have permission to transmit and when you are inside a country's national boundaries, including territorial waters, you are required to operate according to their rules. [*Ham Radio License Manual*, page 7-17]

T1C07 **What may result when correspondence from the FCC is returned as undeliverable because the grantee failed to provide the correct mailing address?**

A. Fine or imprisonment
B. Revocation of the station license or suspension of the operator license
C. Require the licensee to be re-examined
D. A reduction of one rank in operator class

B [97.23] — The FCC requires you to maintain a valid current mailing address in their database at all times. This is so you can be contacted by mail, if needed. If you move or even change PO boxes, be sure to update your information using the FCC ULS online system. If you do not maintain a current address and mail to you is returned to the FCC as undeliverable, your license can be suspended or revoked and removed from the database. [*Ham Radio License Manual*, page 7-8]

T1C08 **What is the normal term for an FCC-issued primary station/ operator license grant?**

A. Five years
B. Life
C. Ten years
D. Twenty years

C [97.25(a)] — The FCC issues all amateur licenses for a 10-year term. You should always renew your license for another 10-year term before it expires. [*Ham Radio License Manual*, page 7-7]

T1C09 **What is the grace period following the expiration of an amateur license within which the license may be renewed?**

A. Two years
B. Three years
C. Five years
D. Ten years

A [97.21(b)] — If you do forget to renew your license, you have up to two years to apply for renewal without having to take the exams again. After the two-year grace period, you will have to retake the exams. Your license is not valid during this two-year grace period, however, and you may not operate an amateur station with an expired license. All that the grace period implies is that the FCC will renew the license if you apply during that period. [*Ham Radio License Manual*, page 7-7]

T1C10 **How soon may you operate a transmitter on an amateur service frequency after you pass the examination required for your first amateur radio license?**

A. Immediately
B. 30 days after the test date
C. As soon as your name and call sign appear in the FCC's ULS database
D. You must wait until you receive your license in the mail from the FCC

C [97.5(a)] — Your Amateur Radio license is valid as soon as the FCC grants the license and posts the information about your license in their database. You don't have to wait for the actual printed license to arrive by mail before you begin to transmit! You can check the FCC database using the FCC Web site or by using an Internet call sign lookup service. The ARRL maintains a call sign lookup service on the ARRL Web site, **www.arrl.org**. [*Ham Radio License Manual*, page 7-7]

T1C11 **If your license has expired and is still within the allowable grace period, may you continue to operate a transmitter on amateur service frequencies?**

A. No, transmitting is not allowed until the ULS database shows that the license has been renewed
B. Yes, but only if you identify using the suffix "GP"
C. Yes, but only during authorized nets
D. Yes, for up to two years

A [97.21(b)] — You may not transmit while your amateur license is expired. You must renew your license before you may transmit. (See also question T1C10.) [*Ham Radio License Manual*, page 7-7]

T1D   Authorized and prohibited transmissions

T1D01 **With which countries are FCC-licensed amateur stations prohibited from exchanging communications?**

A. Any country whose administration has notified the ITU that it objects to such communications
B. Any country whose administration has notified the United Nations that it objects to such communications
C. Any country engaged in hostilities with another country
D. Any country in violation of the War Powers Act of 1934

A [97.111] — You may converse with amateurs in foreign countries unless either amateur's government prohibits the communications. (There are times when a government will not allow its amateurs to talk with people in other countries.) [*Ham Radio License Manual*, page 7-18]

T1D02 **On which of the following occasions may an FCC-licensed amateur station exchange messages with a U.S. military station?**

A. During an Armed Forces Day Communications Test
B. During a Memorial Day Celebration
C. During a Independence Day celebration
D. During a propagation test

A [97.111(a)(5)] — In general, hams can't communicate with non-amateur services, but the FCC may allow hams to talk to non-ham services at certain times or during a declared communications emergency. RACES operators may communicate with government stations during emergencies. The FCC also permits ham-to-military communication on Armed Forces Day. [*Ham Radio License Manual*, page 8-13]

T1D03 When is the transmission of codes or ciphers allowed to hide the meaning of a message transmitted by an amateur station?

A. Only during contests
B. Only when operating mobile
C. Only when transmitting control commands to space stations or radio control craft
D. Only when frequencies above 1280 MHz are used

C [97.113(a)(4), 97.211(b), 97.217] — You can't use codes or ciphers (also known as encryption) to obscure the meaning of transmissions. This means you can't make up a "secret" code to send messages over the air to a friend. However, there are special exceptions. Control signals transmitted for remote control of model craft are not considered codes or ciphers. Neither are telemetry signals, such as a satellite might transmit to report on internal conditions. A space station (satellite) control operator may use specially coded signals to control the satellite. [*Ham Radio License Manual*, page 8-12]

T1D04 What is the only time an amateur station authorized to transmit music?

A. When incidental to an authorized retransmission of manned spacecraft communications
B. When the music produces no spurious emissions
C. When the purpose is to interfere with an illegal transmission
D. When the music is transmitted above 1280 MHz

A [97.113(a)(4), 97.113(e)] — Under FCC Rules, amateurs may not transmit music of any form. Retransmitting music from a broadcast program on a radio in your car or shack is also prohibited — so turn the radio down when you're on the air! There is one exception to the "No Music" rule. If you obtain special permission from NASA to retransmit the audio from a space shuttle mission or the International Space Station for other amateurs to listen, and during that retransmission NASA or the astronauts play some music over the air, that's OK. [*Ham Radio License Manual*, page 8-12]

T1D05 When may amateur radio operators use their stations to notify other amateurs of the availability of equipment for sale or trade?

A. When the equipment is normally used in an amateur station and such activity is not conducted on a regular basis
B. When the asking price is $100.00 or less
C. When the asking price is less than its appraised value
D. When the equipment is not the personal property of either the station licensee or the control operator or their close relatives

A [97.113(a)3] — You are permitted to let other amateurs know that you have Amateur Radio equipment for sale or trade. This is a common practice on local "swap-and-shop" nets. The equipment or materials must be normally used in ham radio! You are also prohibited from selling or trading on a regular basis. [*Ham Radio License Manual*, page 8-11]

T1D06   **Which of the following types of transmissions are prohibited?**
A. Transmissions that contain obscene or indecent words or language
B. Transmissions to establish one-way communications
C. Transmissions to establish model aircraft control
D. Transmissions for third party communications

A   [97.113(a)(4)] — Amateurs may not use obscene or indecent language — it is prohibited by the FCC Rules. Amateur Radio transmissions are public and anyone of any age can hear them. Depending on conditions and frequency, they might be heard anywhere in the world. There is no list of "words you can't say on Amateur Radio", but bear in mind the public nature of our communications and avoid any questionable language. [*Ham Radio License Manual*, page 8-11]

T1D07   **When is an amateur station authorized to automatically retransmit the radio signals of other amateur stations?**
A. When the signals are from an auxiliary, beacon, or Earth station
B. When the signals are from an auxiliary, repeater, or space station
C. When the signals are from a beacon, repeater, or space station
D. When the signals are from an Earth, repeater, or space station

B   [97.113(f)] — Retransmitting the signals of another station is also generally prohibited, except when you are relaying messages or digital data from another station. Some types of stations (repeaters, auxiliary and space stations) are allowed to automatically retransmit signals on different frequencies or channels. [*Ham Radio License Manual*, page 8-12]

T1D08   **When may the control operator of an amateur station receive compensation for operating the station?**
A. When engaging in communications on behalf of their employer
B. When the communication is incidental to classroom instruction at an educational institution
C. When re-broadcasting weather alerts during a RACES net
D. When notifying other amateur operators of the availability for sale or trade of apparatus

B   [97.113] — An exception to the prohibition of being compensated to operate is that teachers may use ham radio as part of their classroom instruction. In that case, they can be a control operator of a ham station, but it must be incidental to their job and can't be the majority of their duties. [*Ham Radio License Manual*, page 8-11]

T1D09 Under which of the following circumstances are amateur stations authorized to transmit signals related to broadcasting, program production, or news gathering, assuming no other means is available?

A. Only where such communications directly relate to the immediate safety of human life or protection of property
B. Only when broadcasting communications to or from the space shuttle.
C. Only where noncommercial programming is gathered and supplied exclusively to the National Public Radio network
D. Only when using amateur repeaters linked to the Internet

A [97.113(b)] — Hams are specifically prohibited from assisting and participating in news gathering by broadcasting organizations, but when the immediate safety of human life or property is involved and normal communications are unavailable, they may use whatever type of communications are necessary. [*Ham Radio License Manual*, page 8-12]

T1D10 What is the meaning of the term broadcasting in the FCC rules for the amateur services?

A. Two-way transmissions by amateur stations
B. Transmission of music
C. Transmission of messages directed only to amateur operators
D. Transmissions intended for reception by the general public

D [97.3(a)(10)] — For the purposes of radio regulation, broadcasting means the transmission of information intended for reception by the general public. These broadcast transmissions may either be direct or relayed. Rules for Amateur Radio, a two-way communications service, prohibit broadcasting by amateur stations. [*Ham Radio License Manual*, page 8-12]

T1D11 Which of the following types of communications are permitted in the Amateur Radio Service?

A. Brief transmissions to make station adjustments
B. Retransmission of entertainment programming from a commercial radio or TV station
C. Retransmission of entertainment material from a public radio or TV station
D. Communications on a regular basis that could reasonably be furnished alternatively through other radio services

A [97.113(a)(5)] — If you need to make a short transmission to test an antenna or make adjustments to your radio, make sure you identify your station by sending or speaking your call sign. [*Ham Radio License Manual*, page 8-3]

T1E Control operator and control types; control operator required, eligibility, designation of control operator, privileges and duties, control point, local, automatic and remote control, location of control operator

T1E01   When must an amateur station have a control operator?

A. Only when the station is transmitting
B. Only when the station is being locally controlled
C. Only when the station is being remotely controlled
D. Only when the station is being automatically controlled

A   [97.7(a)] — All amateur transmissions must be made under the supervision of a control operator whose job it is to be sure FCC Rules are followed. This is true even if the station is operating under automatic control. [*Ham Radio License Manual*, page 8-1]

T1E02   Who is eligible to be the control operator of an amateur station?

A. Only a person holding an amateur service license from any country that belongs to the United Nations
B. Only a citizen of the United States
C. Only a person over the age of 18
D. Only a person for whom an amateur operator/primary station license grant appears in the FCC database or who is authorized for alien reciprocal operation

D   [97.7(a)] — A control operator must be named in the FCC amateur license database or be an alien with reciprocal operating authorization. (An alien is a citizen of another country.) This is a simple requirement — the FCC has to know who you are, that you are licensed, and where you can be contacted. Any licensed amateur can be a control operator. [*Ham Radio License Manual*, page 8-2]

T1E03   Who must designate the station control operator?

A. The station licensee
B. The FCC
C. The frequency coordinator
D. The ITU

A   [97.103(b)] — The control operator doesn't have to be the station licensee and doesn't even have to be physically present at the transmitter, but all amateur transmissions are the responsibility of a control operator. The station licensee is responsible for designating the control operator. [*Ham Radio License Manual*, page 8-1]

T1E04   **What determines the transmitting privileges of an amateur station?**

A. The frequency authorized by the frequency coordinator
B. The class of operator license held by the station licensee
C. The highest class of operator license held by anyone on the premises
D. The class of operator license held by the control operator

D   [97.103(b)] — An amateur station may transmit in any way permitted by the privileges of the control operator's license class. [*Ham Radio License Manual*, page 8-2]

T1E05   **What is an amateur station control point?**

A. The location of the station's transmitting antenna
B. The location of the station transmitting apparatus
C. The location at which the control operator function is performed
D. The mailing address of the station licensee

C   [97.3(a)(14)] — This sounds like a circular definition, but the rule allows for an amateur station to be controlled from a remote location or even under automated control by devices and procedures that ensure proper operation. [*Ham Radio License Manual*, page 8-2]

T1E06   **Under which of the following types of control is it permissible for the control operator to be at a location other than the control point?**

A. Local control
B. Automatic control
C. Remote control
D. Indirect control

B   [97.109(d)] — Under automatic control, the station operates completely under the control of devices and procedures that ensure compliance with FCC rules. A control operator is still required, but need not be at the control point when the station is transmitting. [*Ham Radio License Manual*, page 8-10]

T1E07   **When the control operator is not the station licensee, who is responsible for the proper operation of the station?**

A. All licensed amateurs who are present at the operation
B. Only the station licensee
C. Only the control operator
D. The control operator and the station licensee are equally responsible

D   [97.103(a)] — Regardless of license class, both the control operator and station owner are responsible for proper operation of the station. The control operator is responsible for any transmissions made by the station and the licensee is responsible for controlling access to the station. [*Ham Radio License Manual*, page 8-2]

T1E08 What type of control is being used for a repeater when the control operator is not present at a control point?

A. Local control
B. Remote control
C. Automatic control
D. Unattended

C [97.3(a)] — See question T1E06. [*Ham Radio License Manual*, page 8-10]

T1E09 What type of control is being used when transmitting using a handheld radio?

A. Radio control
B. Unattended control
C. Automatic control
D. Local control

D [97.109(a)] — Because the operator is directly manipulating the transmitter, this is local control. [*Ham Radio License Manual*, page 8-10]

T1E10 What type of control is used when the control operator is not at the station location but can indirectly manipulate the operating adjustments of a station?

A. Local
B. Remote
C. Automatic
D. Unattended

B [97.3] — Because the transmitter is manipulated by the control operator through a control link, this is remote control. [*Ham Radio License Manual*, page 8-10]

T1E11 Who does the FCC presume to be the control operator of an amateur station, unless documentation to the contrary is in the station records?

A. The station custodian
B. The third party participant
C. The person operating the station equipment
D. The station licensee

D [97.103(a)] — The FCC will presume the station licensee to be the control operator unless there is a written record to the contrary. This is another reason why you should keep a logbook. [*Ham Radio License Manual*, page 8-2]

T1F Station identification and operation standards; special operations for repeaters and auxiliary stations, third party communications, club stations, station security, FCC inspection

T1F01 What type of identification is being used when identifying a station on the air as "Race Headquarters"?

A. Tactical call
B. Self-assigned designator
C. SSID
D. Broadcast station

A   Tactical call signs (or tactical IDs) are used to help identify where a station is and what it is doing. Examples of tactical calls include "Waypoint 5," "First Aid Station," "Hollywood and Vine" and "Fire Watch on Coldwater Ridge." [*Ham Radio License Manual*, page 8-3]

T1F02 When using tactical identifiers, how often must you transmit the station's FCC-assigned call sign?

A. Never, the tactical call is sufficient
B. Once during every hour
C. Every ten minutes
D. At the end of every communication

C   [97.119(a)] — In addition to using the tactical identifier, identification using the station's FCC-assigned call sign is still required as stated in the FCC Rules. (See question T1F03.) [*Ham Radio License Manual*, page 8-4]

T1F03 When is an amateur station required to transmit its assigned call sign?

A. At the beginning of each contact, and every 10 minutes thereafter
B. At least once during each transmission
C. At least every 15 minutes during and at the end of a contact
D. At least every 10 minutes during and at the end of a contact

D   [97.119(a)] — FCC regulations are very specific about station identification. You must identify your station every 10 minutes (or more frequently) during a contact and at the end of the contact. [*Ham Radio License Manual*, page 8-3]

T1F04 **Which of the following is an acceptable language for use for station identification when operating in a phone sub-band?**

A. Any language recognized by the United Nations
B. Any language recognized by the ITU
C. The English language
D. English, French, or Spanish

C [97.119(b)(2)] — You can use any language you want to communicate with other amateurs, even Esperanto! Amateur Radio provides you a great opportunity to practice your foreign language skills with other hams that speak that language. When you give your identification as a US station, however, you must use English. Foreign stations may have a similar requirement to use their native language when identifying their transmissions. [*Ham Radio License Manual*, page 8-3]

T1F05 **What method of call sign identification is required for a station transmitting phone signals?**

A. Send the call sign followed by the indicator RPT
B. Send the call sign using CW or phone emission
C. Send the call sign followed by the indicator R
D. Send the call sign using only phone emission

B [97.119(b)] — The call sign ID must be transmitted with an emission type authorized for the transmitting frequency, so station transmitting phone signals must identify its transmissions by means of spoken call signs or Morse code. The sending speed of a CW ID must not exceed 20 WPM (words per minute) if keyed by an automatic device used only for identification, a common practice for repeater stations. [*Ham Radio License Manual*, page 8-3]

T1F06 **Which of the following formats of a self-assigned indicator is acceptable when identifying using a phone transmission?**

A. KL7CC stroke W3
B. KL7CC slant W3
C. KL7CC slash W3
D. All of these choices are correct

D [97.119(c)] — An indicator is words, letters or numerals appended to and separated from your call sign during identification. When operating away from your home station, you should add an indicator to your call sign so that other stations are aware of your location. For instance, you might add the word "mobile" to your call sign when you operating from your car on your way to work. This is important if using your regular call sign would be confusing. If Alaska resident KL7CC is operating from a location in the 3rd district, he would give his call sign as "KL7CC/W3." The added "/W3" is a self-assigned indicator. [*Ham Radio License Manual*, page 8-4]

T1F07   **Which of the following restrictions apply when appending a self-assigned call sign indicator?**

A. It must be more than three letters and less than five letters
B. It must be less than five letters
C. It must start with the letters AA through AL, K, N, or W and be not less than two characters or more than five characters in length
D. It must not conflict with any other indicator specified by the FCC rules or with any call sign prefix assigned to another country

D   [97.119(c)] — A self-assigned indicator is not allowed to conflict with some other indicator specified in the FCC rules, such as AG or AE that indicate an upgraded license class. It may not conflict with a prefix assigned to some other country. For example, adding the indicator /M to indicate mobile operation is not allowed because M is an English prefix, but /MOB is permitted because MOB is not a prefix from another country. [*Ham Radio License Manual*, page 8-4]

T1F08   **When may a Technician Class licensee be the control operator of a station operating in an exclusive Extra Class segment of the amateur bands?**

A. Never
B. On Armed Forces Day
C. As part of a multi-operator contest team
D. When using a club station whose trustee is an Extra Class operator licensee

A   [97.105(b)] — If you are the control operator of a station, you may only operate according to your license privileges. (See question T1E04.) [*Ham Radio License Manual*, page 8-2]

T1F09   **What type of amateur station simultaneously retransmits the signal of another amateur station on a different channel or channels?**

A. Beacon station
B. Earth station
C. Repeater station
D. Message forwarding station

C   [97.3(a)(39)] — See question T1D07. [*Ham Radio License Manual*, page 2-11]

T1F10 Who is accountable should a repeater inadvertently retransmit communications that violate the FCC rules?

A. The control operator of the originating station
B. The control operator of the repeater
C. The owner of the repeater
D. Both the originating station and the repeater owner

A [97.205(g)] — Although there must be a control operator responsible for proper operation of the repeater, the primary responsibility for complying with FCC Rules for operating lies with the transmitting station. If repeater users consistently violate operating rules, the FCC can require that a repeater system be placed on remote control, meaning a control operator must monitor repeater operation. [*Ham Radio License Manual*, page 8-10]

T1F11 To which foreign stations do the FCC rules authorize the transmission of non-emergency third party communications?

A. Any station whose government permits such communications
B. Those in ITU Region 2 only
C. Those in ITU Regions 2 and 3 only
D. Those in ITU Region 3 only

A [97.115(a)] — There must be a third-party agreement in place between both governments for third-party communications to be permitted. [*Ham Radio License Manual*, page 8-9]

T1F12 How many persons are required to be members of a club for a club station license to be issued by the FCC?

A. At least 5
B. At least 4
C. A trustee and 2 officers
D. At least 2

B [97.5(b)(2)] — In order to ensure that a club actually exists, the FCC requires that it have at least four members, a name, be formally organized and managed and have a primary purpose devoted to Amateur Radio. [*Ham Radio License Manual*, page 7-21]

T1F13 **When must the station licensee make the station and its records available for FCC inspection?**

A. Any time upon request by an official observer
B. Any time upon request by an FCC representative
C. 30 days prior to renewal of the station license
D. 10 days before the first transmission

B [97.103(c)] — As a station licensee, your responsibilities include the requirement to make the station and the station records available for inspection upon request by an FCC representative. An inspection can occur at any time, although they are actually quite rare. [*Ham Radio License Manual*, page 7-8]

Operating Procedures

SUBELEMENT T2 — Operating Procedures
[3 Exam Questions — 3 Groups]

T2A   Station operation; choosing an operating frequency, calling another station, test transmissions, use of minimum power, frequency use, band plans

T2A01   What is the most common repeater frequency offset in the 2 meter band?

A. Plus 500 kHz
B. Plus or minus 600 kHz
C. Minus 500 kHz
D. Only plus 600 kHz

B   The amount of repeater offset or shift is the same for almost all repeaters on one band as shown in the table below. This standardization makes it easy to use many different repeaters. Note that offset can often be either positive or negative by region or to allow more repeaters to occupy a single band, so be sure to use a repeater directory to determine the proper offset. [*Ham Radio License Manual*, page 6-16]

Standard Repeater Offsets by Band

Band	Offset
10 meters	−100 kHz
6 meters	Varies by region: −500 kHz, −1 MHz, −1.7 MHz
2 meters	+ or −600 kHz
1.25 meters	−1.6 MHz
70 cm	+ or −5 MHz
902 MHz	12 MHz
1296 MHz	12 MHz

T2A02   What is the national calling frequency for FM simplex operations in the 70 cm band?

A. 146.520 MHz
B. 145.000 MHz
C. 432.100 MHz
D. 446.000 MHz

D   446.000 MHz — the nationwide calling frequencies for FM simplex are listed in the table below. [*Ham Radio License Manual*, page 6-14]

FM Simplex Calling Frequencies

Band	Frequency (MHz)
10 meters	29.600
6 meters	52.525
2 meters	146.52
1.25 meters	223.50
70 cm	446.00
902 MHz	906.50
1296 MHz	1294.50

T2A03   What is a common repeater frequency offset in the 70 cm band?

A. Plus or minus 5 MHz
B. Plus or minus 600 kHz
C. Minus 600 kHz
D. Plus 600 kHz

A   See question T2A02. [*Ham Radio License Manual*, page 6-16]

T2A04   What is an appropriate way to call another station on a repeater if you know the other station's call sign?

A. Say "break, break" then say the station's call sign
B. Say the station's call sign then identify with your call sign
C. Say "CQ" three times then the other station's call sign
D. Wait for the station to call "CQ" then answer it

B   To call another station when the repeater is not in use, just give both call signs. For example, "N1IL, this is N1BKE." If the repeater is in use, but the conversation sounds like it is about to end, wait before calling another station. [*Ham Radio License Manual*, page 6-12]

T2A05   What should you transmit when responding to a call of CQ?

A. CQ followed by the other station's call sign
B. Your call sign followed by the other station's call sign
C. The other station's call sign followed by your call sign
D. A signal report followed by your call sign

C   On phone, say the other station's call sign followed by "this is" and your call sign at least once using phonetics, such as those in the table of ITU phonetics below. On CW, substitute the abbreviation DE for "this is." There is no need to send your call sign more than twice unless there is a lot of noise or interference present or the calling station is very weak. [*Ham Radio License Manual*, page 6-13]

ITU Phonetic Alphabet

Letter	Word	Pronunciation
A	Alfa	**AL** FAH
B	Bravo	**BRAH** VOH
C	Charlie	**CHAR** LEE
D	Delta	**DELL** TAH
E	Echo	**ECK** OH
F	Foxtrot	**FOKS** TROT
G	Golf	GOLF
H	Hotel	HOH **TELL**
I	India	**IN** DEE AH
J	Juliet	**JEW** LEE ETT
K	Kilo	**KEY** LOH
L	Lima	**LEE** MAH
M	Mike	MIKE
N	November	NO **VEM** BER
O	Oscar	**OSS** CAH
P	Papa	PAH **PAH**
Q	Quebec	KEH **BECK**
R	Romeo	**ROW** ME OH
S	Sierra	SEE **AIR** RAH
T	Tango	**TANG** GO
U	Uniform	**YOU** NEE FORM
V	Victor	**VIK** TAH
W	Whiskey	**WISS** KEY
X	X-Ray	**ECKS** RAY
Y	Yankee	**YANG** KEY
Z	Zulu	**ZOO** LOO

Note: The **boldfaced** syllables are emphasized. The pronunciations shown in this table were designed for those who speak any of the international languages. The pronunciations given for "Oscar" and "Victor" may seem awkward to English-speaking people in the US.

T2A06   **What must an amateur operator do when making on-air transmissions to test equipment or antennas?**

A. Properly identify the transmitting station
B. Make test transmissions only after 10:00 p.m. local time
C. Notify the FCC of the test transmission
D. State the purpose of the test during the test procedure

A   Unidentified transmissions are not allowed, so be sure to follow FCC Rules. The identification rules are simple — give your call sign at least once every 10 minutes during a contact and when the contact is finished. To establish contact you will need to give your call sign as well, although it's not required. If the transmission or contact is too short for the 10 minute rule, just give your call sign as you end the contact. [*Ham Radio License Manual*, page 8-3]

T2A07   **Which of the following is true when making a test transmission?**

A. Station identification is not required if the transmission is less than 15 seconds
B. Station identification is not required if the transmission is less than 1 watt
C. Station identification is required only if your station can be heard
D. Station identification is required at least every ten minutes during the test and at the end

D   See question T2A06. [*Ham Radio License Manual*, page 8-3]

T2A08   **What is the meaning of the procedural signal "CQ"?**

A. Call on the quarter hour
B. A new antenna is being tested (no station should answer)
C. Only the called station should transmit
D. Calling any station

D   CQ literally means "Calling any station." You can usually tell good operators by how they call CQ. A good operator makes short, crisp calls separated by listening periods. Think of a CQ as an advertisement for your station and your operating skills. [*Ham Radio License Manual*, page 6-13]

T2A09 **What brief statement is often used in place of "CQ" to indicate that you are listening on a repeater?**

A. Say "Hello test" followed by your call sign
B. Say your call sign
C. Say the repeater call sign followed by your call sign
D. Say the letters "QSY" followed by your call sign

B Since a repeater's signal is generally strong and the output frequency fixed, there is no need for an extended CQ call as there is when using SSB or CW. All that's necessary is to announce your presence and anyone who wishes to call you will then do so. It is also common to say "monitoring" following your call sign to reinforce that you are present and listening. [*Ham Radio License Manual*, page 6-12]

T2A10 **What is a band plan, beyond the privileges established by the FCC?**

A. A voluntary guideline for using different modes or activities within an amateur band
B. A mandated list of operating schedules
C. A list of scheduled net frequencies
D. A plan devised by a club to use a frequency band during a contest

A Band plans are voluntary agreements between operators about how to use the bands under normal circumstances. These plans go into more detail on using different operating modes within an amateur band than is specified in FCC regulations. Good operators are familiar with the band plans and try to follow them. [*Ham Radio License Manual*, page 7-15]

T2A11 **What are the FCC rules regarding power levels used in the amateur bands?**

A. Always use the maximum power allowed to ensure that you complete the contact
B. An amateur may use no more than 200 watts PEP to make an amateur contact
C. An amateur may use up to 1500 watts PEP on any amateur frequency
D. An amateur must use the minimum transmitter power necessary to carry out the desired communication

D [97.313(a)] — Although the maximum power level available to amateurs is 1500 watts PEP, the vast majority of contacts require far less power. Amateurs are required to avoid using excessive power levels to allow more hams to use the frequency. This does not mean you must reduce power until the other operator is barely able to hear you, just use a power level that provides satisfactory results. [*Ham Radio License Manual*, page 7-14]

T2B — VHF/UHF operating practices; SSB phone, FM repeater, simplex, frequency offsets, splits and shifts, CTCSS, DTMF, tone squelch, carrier squelch, phonetics

T2B01 What is the term used to describe an amateur station that is transmitting and receiving on the same frequency?

A. Full duplex communication
B. Diplex communication
C. Simplex communication
D. Half duplex communication

C In Amateur Radio, simplex operation means that the stations are communicating by transmitting and receiving on the same frequency. Using a repeater with different transmit and receive frequencies is called *duplex* operation. [*Ham Radio License Manual*, page 6-10]

T2B02 What is the term used to describe the use of a sub-audible tone transmitted with normal voice audio to open the squelch of a receiver?

A. Carrier squelch
B. Tone burst
C. DTMF
D. CTCSS

D Repeater access tones were invented by Motorola to allow different commercial users to share a repeater without having to listen to each other's conversations. These tones are known by various names: Continuous Tone Coded Squelch System (CTCSS), PL (for Private Line, the Motorola trade name), or subaudible. FRS/GMRS radio users know these tones as privacy codes or privacy tones. [*Ham Radio License Manual*, page 6-16]

T2B03 Which of the following describes the muting of receiver audio controlled solely by the presence or absence of an RF signal?

A. Tone squelch
B. Carrier squelch
C. CTCSS
D. Modulated carrier

B To keep from having to listen to continuous noise when no signal is present, the squelch circuit was invented. The squelch circuit (sometimes called *carrier squelch*) mutes the receiver's audio output when no signal is present. [*Ham Radio License Manual*, page 5-6]

T2B04 **Which of the following common problems might cause you to be able to hear but not access a repeater even when transmitting with the proper offset?**

A. The repeater receiver requires audio tone burst for access
B. The repeater receiver requires a CTCSS tone for access
C. The repeater receiver may require a DCS tone sequence for access
D. All of these choices are correct

D If you can hear a repeater's signal and you're sure you are using the right offset, but you can't access the repeater, then you probably don't have your radio set up to use the right type or frequency of access tone. [*Ham Radio License Manual*, page 6-17]

T2B05 **What determines the amount of deviation of an FM signal?**

A. Both the frequency and amplitude of the modulating signal
B. The frequency of the modulating signal
C. The amplitude of the modulating signal
D. The relative phase of the modulating signal and the carrier

C The frequency of an FM signal varies with the amplitude of the modulating signal as shown in the figure below. The amount of variation is called *carrier deviation* or just deviation. [*Ham Radio License Manual*, page 2-9]

Figure T2-1 — At (A), each cycle of the unmodulated carrier has the same frequency. When the carrier is frequency-modulated with the signal at (B), its frequency increases and decreases corresponding to the increases and decreases in amplitude of the modulating signal.

T2B06 **What happens when the deviation of an FM transmitter is increased?**

A. Its signal occupies more bandwidth
B. Its output power increases
C. Its output power and bandwidth increases
D. Asymmetric modulation occurs

A (See also question T2B05.) As deviation increases, so does the signal's bandwidth, so excessive deviation can cause interference to nearby signals. [*Ham Radio License Manual*, page 2-9]

T2B07 **What should you do if you receive a report that your station's transmissions are causing splatter or interference on nearby frequencies?**

A. Increase transmit power

B. Change mode of transmission

C. Report the interference to the equipment manufacturer

D. Check your transmitter for off-frequency operation or spurious emissions

D Your equipment can cause spurious emissions if you operate it with some controls adjusted improperly. For example, if you set your microphone gain too high on an SSB transmitter, the resulting overmodulation creates spurious signals on nearby frequencies. On FM, misadjusting your tuning control or speaking too loudly (causing excessive deviation) can cause your signal to encroach on an adjacent channel. Every ham should make sure to transmit in a way that minimizes the possibility of causing harmful interference. Reports of interference such as transmitting off-frequency or generating spurious signals (splatter and buckshot) should always be checked out. [*Ham Radio License Manual*, page 8-7]

T2B08 **What is the proper course of action if your station's transmission unintentionally interferes with another station?**

A. Rotate your antenna slightly

B. Properly identify your transmission and move to a different frequency

C. Increase power

D. Change antenna polarization

B Accidental interference happens frequently — perhaps you didn't hear the other stations before transmitting. Simple courtesy works wonders! Just say "excuse me," give your call sign, and move to a different frequency. [*Ham Radio License Manual*, page 8-8]

T2B09 **Which of the following methods is encouraged by the FCC when identifying your station when using phone?**

A. Use of a phonetic alphabet

B. Send your call sign in CW as well as voice

C. Repeat your call sign three times

D. Increase your signal to full power when identifying

A [97.119(b)(2)] — If the other operator is having difficulty copying your signals you should use the standard International Telecommunication Union (ITU) phonetic alphabet (see question T2A05). Use the words in the phonetic alphabet to spell out the letters in your call sign, your name or any other piece of information that might be confused if the letters are not received correctly. This phonetic alphabet is generally understood by hams in all countries. [*Ham Radio License Manual*, page 8-3]

T2B10 What is the "Q" signal used to indicate that you are receiving interference from other stations?

A. QRM
B. QRN
C. QTH
D. QSB

A Q-signals are a system of making queries and exchanging information in an abbreviated form. They also allow operators who speak different languages to communicate. QRM refers to interference from other stations. QRN refers to interference from atmospheric static, and QTH means the station's location. QSB indicates signal fading. There are many useful Q-signals used by amateurs. [*Ham Radio License Manual*, page 6-5]

T2B11 What is the "Q" signal used to indicate that you are changing frequency?

A. QRU
B. QSY
C. QSL
D. QRZ

B QSY means, "I am changing frequency." (See also question T2B10.) [*Ham Radio License Manual*, page 6-5]

T2C — Public service; emergency and non-emergency operations, message traffic handling

T2C01 What set of rules applies to proper operation of your station when using amateur radio at the request of public service officials?

A. RACES Rules
B. ARES Rules
C. FCC Rules
D. FEMA Rules

C [97.103(a)] — At all times, you are bound by FCC rules, even if using your radio in support of a public safety agency. If you are requested to use the station in a way that violates the FCC rules (except in the case of there being an immediate threat to life or property), you should politely decline. [*Ham Radio License Manual*, page 6-26]

T2C02 Who must submit the request for a temporary waiver of Part 97.113 to allow amateur radio operators to provide communications on behalf of their employers during a government sponsored disaster drill?

A. Each amateur participating in the drill
B. Any employer participating in the drill
C. The local American Red Cross Chapter
D. The government agency sponsoring the event

D [97.113 and FCC Public Notice DA 09-2259] — Operating on behalf of or at the direction of your employer, even when off-duty, is prohibited. Participating in training exercises and drills organized by your employer is allowed only if two conditions are met: The drill must be government-sponsored and your employer is a government agency that has obtained a written waiver for the exercise from the FCC. [*Ham Radio License Manual*, page 6-25]

 T2C03 When is it legal for an amateur licensee to provide communications on behalf of their employer during a government sponsored disaster drill or exercise?

A. Whenever the employer is a not-for-profit organization
B. Whenever there is a temporary need for the employer's business continuity plan
C. Only when the FCC has granted a government-requested waiver
D. Only when the amateur is not receiving compensation from his employer for the activity

C [97.113] — See question T2C02. [*Ham Radio License Manual*, page 6-25]

T2C04 What do RACES and ARES have in common?

A. They represent the two largest ham clubs in the United States
B. Both organizations broadcast road and weather traffic information
C. Neither may handle emergency traffic supporting public service agencies
D. Both organizations may provide communications during emergencies

D RACES stands for Radio Amateur Civil Emergency Service, a communications service within the amateur service. RACES provides communications assistance to civil defense organizations in times of need. It is active only during periods of local, regional or national civil emergencies. ARES stands for Amateur Radio Emergency Service and is sponsored by the ARRL. ARES presents a way for local amateurs to provide emergency communications while working with local public safety agencies and groups such as the Red Cross. ARES can provide communications assistance at any time. [*Ham Radio License Manual*, page 6-27]

T2C05 What is the Radio Amateur Civil Emergency Service?

A. An emergency radio service organized by amateur operators
B. A radio service using amateur stations for emergency management or civil defense communications
C. A radio service organized to provide communications at civic events
D. A radio service organized by amateur operators to assist non-military persons

B [97.3(a)(37), 97.407] — See question T2C04. [*Ham Radio License Manual*, page 6-27]

T2C06 Which of the following is common practice during net operations to get the immediate attention of the net control station when reporting an emergency?

A. Repeat the words SOS three times followed by the call sign of the reporting station
B. Press the push-to-talk button three times
C. Begin your transmission with "Priority" or "Emergency" followed by your call sign
D. Play a pre-recorded emergency alert tone followed by your call sign

C It is important to remember that no matter what the purpose or status of a net, a station with emergency traffic should break in at any time. If the net is operating on phone and you are reporting an emergency, break in by saying "Priority" or "Emergency," followed by your call sign. [*Ham Radio License Manual*, page 6-22]

T2C07 What should you do to minimize disruptions to an emergency traffic net once you have checked in?

A. Whenever the net frequency is quiet, announce your call sign and location
B. Move 5 kHz away from the net's frequency and use high power to ask other hams to keep clear of the net frequency
C. Do not transmit on the net frequency until asked to do so by the net control station
D. Wait until the net frequency is quiet, then ask for any emergency traffic for your area

C The desire to help may be strong, but remember that unnecessary transmissions just slow things down. The Net Control Station (NCS) will record your call sign and location so that if you're needed, you can be called. There's no need to remind the NCS that you're listening. Once you've checked in, you should not transmit unless you are specifically requested or authorized to do so or a request is made for capabilities or information that you can provide. [*Ham Radio License Manual*, page 6-22]

T2C08 What is usually considered to be the most important job of an amateur operator when handling emergency traffic messages?

A. Passing messages exactly as written, spoken or as received
B. Estimating the number of people affected by the disaster
C. Communicating messages to the news media for broadcast outside the disaster area
D. Broadcasting emergency information to the general public

A The most important job during emergency and disaster net operation is the ability to accurately relay or "pass" messages exactly as written, spoken or received. [*Ham Radio License Manual*, page 6-22]

T2C09 When may an amateur station use any means of radio communications at its disposal for essential communications in connection with immediate safety of human life and protection of property?

A. Only when FEMA authorizes it by declaring an emergency
B. When normal communications systems are not available
C. Only when RACES authorizes it by declaring an emergency
D. Only when authorized by the local MARS program director

B [97.403] — Stations in distress should use any means at their disposal to attract attention and obtain assistance. This includes operating on frequencies and equipment for which they are not licensed. In an emergency situation where there is immediate risk to life or property and normal forms of communication are unavailable, you may use any means possible to address that risk. [*Ham Radio License Manual*, page 6-26]

T2C10   What is the preamble in a formal traffic message?

A. The first paragraph of the message text
B. The message number
C. The priority handling indicator for the message
D. The information needed to track the message as it passes through the amateur radio traffic handling system

D   Each message needs some information about the message and its content. The preamble gives the message a unique identity. The preamble consists of the following information:

Number — a unique number assigned by the station that creates the radiogram

Precedence — a description of the nature of the radiogram; Routine, Priority, Emergency and Welfare

Handling Instructions (HX) — for special instructions in how to the handle the radiogram.

Station of Origin — the call sign of the radio station from which the radiogram was first sent by Amateur Radio. (This allows information about the message to be returned to the sending station.)

Check — a count of the number of words or word equivalents in the text of the radiogram.

Place of Origin — the name of the town from which the radiogram started.

Time and Date — the time and date the radiogram is received at the station that first sent it.

Address — the complete name, street and number, city and state where the radiogram is going

[*Ham Radio License Manual*, page 6-22]

T2C11   What is meant by the term "check" in reference to a formal traffic message?

A. The check is a count of the number of words or word equivalents in the text portion of the message
B. The check is the value of a money order attached to the message
C. The check is a list of stations that have relayed the message
D. The check is a box on the message form that tells you the message was received

A   See question T2C10. [*Ham Radio License Manual*, page 6-22]

Radio Waves, Propagation and Antennas

SUBELEMENT T3 — Radio wave characteristics, radio and electromagnetic properties, propagation modes [3 Exam Questions — 3 Groups]

T3A   Radio wave characteristics; how a radio signal travels; distinctions of HF, VHF and UHF; fading, multipath; wavelength vs. penetration; antenna orientation

T3A01   What should you do if another operator reports that your station's 2 meter signals were strong just a moment ago, but now they are weak or distorted?

A. Change the batteries in your radio to a different type
B. Turn on the CTCSS tone
C. Ask the other operator to adjust his squelch control
D. Try moving a few feet, as random reflections may be causing multipath distortion

D   In a location where reflections can occur, such as near vehicles or buildings, it is possible for a direct signal and a reflected signal to interfere with each other at the receiver and partially cancel. When this occurs, all that is necessary is to move approximately one-half wavelength in any direction — about three feet on the 2 meter band. It is likely that you'll find a "hot spot" within that range, or at least a location where the cancellation is greatly reduced. [*Ham Radio License Manual*, page 4-2]

T3A02   Why are UHF signals often more effective from inside
buildings than VHF signals?

A. VHF signals lose power faster over distance
B. The shorter wavelength allows them to more easily penetrate the
structure of buildings
C. This is incorrect; VHF works better than UHF inside buildings
D. UHF antennas are more efficient than VHF antennas

B   Radio waves can penetrate openings in otherwise solid objects as long as
at least one side of the opening is longer than about one-half wavelength. For
this reason, the shorter wavelengths of UHF signals make them more effective
at propagating into and out of buildings in urban areas. Larger obstructions
also diffract or bend signals better, so UHF signals also have less shadowing
effects in urban areas. [*Ham Radio License Manual*, page 4-1]

T3A03   What antenna polarization is normally used for long-distance
weak-signal CW and SSB contacts using the VHF and UHF
bands?

A. Right-hand circular
B. Left-hand circular
C. Horizontal
D. Vertical

C   Horizontally polarized Yagis and quads are usually used for long-
distance communications, especially for "weak signal" SSB and CW contacts
on the VHF and UHF bands. Horizontal polarization is preferred because it
results in lower ground losses when the wave reflects from or travels along
the ground. [*Ham Radio License Manual*, page 4-15]

T3A04   What can happen if the antennas at opposite ends of a VHF or
UHF line of sight radio link are not using the same
polarization?

A. The modulation sidebands might become inverted
B. Signals could be significantly weaker
C. Signals have an echo effect on voices
D. Nothing significant will happen

B   On the VHF and UHF bands, it is important to keep the transmitting
and receiving antennas aligned so that they have matching polarizations.
If the radio wave from a transmitter arrives at the receiver with a different
polarization, the receiving antenna does not respond as well to the incoming
radio wave. Since most repeaters have vertically polarized antennas, when
using a low-power handheld radio to access a repeater it is important that
your antenna be vertically polarized, too. [*Ham Radio License Manual*,
page 4-6]

T3A05 When using a directional antenna, how might your station be able to access a distant repeater if buildings or obstructions are blocking the direct line of sight path?

A. Change from vertical to horizontal polarization
B. Try to find a path that reflects signals to the repeater
C. Try the long path
D. Increase the antenna SWR

B Remember that buildings and hills can act as radio reflectors. If you can "aim" your signal at one of these large reflectors, it is often possible for your signal to avoid an obstruction directly in your line of sight to the desired station. [*Ham Radio License Manual*, page 4-14]

T3A06 What term is commonly used to describe the rapid fluttering sound sometimes heard from mobile stations that are moving while transmitting?

A. Flip-flopping
B. Picket fencing
C. Frequency shifting
D. Pulsing

B Imagine yourself driving through an area with lots of reflection. You will travel in and out of "hot spots" and "dead spots" quite rapidly. The effect on your signal is an equally rapid increase and decrease in signal strength called mobile flutter or picket fencing. [*Ham Radio License Manual*, page 4-2]

T3A07 What type of wave carries radio signals between transmitting and receiving stations?

A. Electromagnetic
B. Electrostatic
C. Surface acoustic
D. Magnetostrictive

A An electromagnetic wave is a combination of an electric and a magnetic field, just like the fields in a capacitor and inductor but spreading out into space like ripples traveling across the surface of water. The wave's electric and magnetic fields oscillate at the same frequency as the RF current in the transmitting antenna. [*Ham Radio License Manual*, page 4-6]

T3A08   What is the cause of irregular fading of signals from distant stations during times of generally good reception?

A. Absorption of signals by the "D" layer of the ionosphere
B. Absorption of signals by the "E" layer of the ionosphere
C. Random combining of signals arriving via different path lengths
D. Intermodulation distortion in the local receiver

C   Any time signals can take more than one path from the transmitter to the receiver, it is possible that they will partially cancel, causing the combined signals to fade out. This can occur on any band, although it is more common on the HF bands because of variations in the ionosphere. [*Ham Radio License Manual*, page 4-2]

T3A09   Which of the following is a common effect of "skip" reflections between the Earth and the ionosphere?

A. The sidebands become reversed at each reflection
B. The polarization of the original signal is randomized
C. The apparent frequency of the received signal is shifted by a random amount
D. Signals at frequencies above 30 MHz become stronger with each reflection

B   The characteristics of the ionosphere vary with solar radiation, whether it is day or night, or from different concentrations of molecules and ions. As a result, as a radio wave travels through the ionosphere its polarization often changes randomly so that when received, it can have any polarization. [*Ham Radio License Manual*, page 4-6]

T3A10   What may occur if VHF or UHF data signals propagate over multiple paths?

A. Transmission rates can be increased by a factor equal to the number of separate paths observed
B. Transmission rates must be decreased by a factor equal to the number of separate paths observed
C. No significant changes will occur if the signals are transmitting using FM
D. Error rates are likely to increase

D   Variations in signal strength from multipath that cause fading of voice signals can also cause digital data signals to be received with a higher error rate, particularly at VHF and UHF. (See question T3A06.) [*Ham Radio License Manual*, page 4-2]

T3A11 Which part of the atmosphere enables the propagation of radio signals around the world?

A. The stratosphere
B. The troposphere
C. The ionosphere
D. The magnetosphere

C Radio waves at HF (and sometimes VHF) can be completely bent back towards the earth by diffraction in the ionosphere's F-layers as if they were reflected. This is called sky wave propagation or skip. Since the Earth's surface is also conductive, it can also reflect radio waves. This means that a radio wave can be reflected between the ionosphere and ground multiple times. Each reflection from the ionosphere is called a hop and allows radio waves to be received hundreds or thousands of miles away. This is the most common way for hams to make long-distance contacts on the HF bands. [*Ham Radio License Manual*, page 4-3]

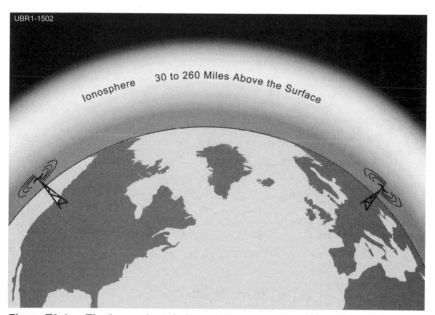

Figure T3-1 — The ionosphere is formed by solar ultraviolet (UV) radiation. The UV rays knock electrons loose from air molecules, creating weakly charged layers at different heights. These layers can absorb or refract radio signals, sometimes bending them back to the Earth.

T3B Radio and electromagnetic wave properties; the electromagnetic spectrum, wavelength vs. frequency, velocity of electromagnetic waves

T3B01 What is the name for the distance a radio wave travels during one complete cycle?

A. Wave speed
B. Waveform
C. Wavelength
D. Wave spread

C A radio wave travels at a constant speed, the speed of light. That means the radio wave will always cover the same distance during the time it takes the wave's oscillating electric and magnetic fields to make one complete cycle. That distance is the wave's wavelength. All radio waves of the same frequency, traveling at the speed of light, will have the same wavelength. [*Ham Radio License Manual*, page 2-4]

Radio wave oscillating at a frequency = f

ARRL0013

DISTANCE

Distance the wave travels during one cycle = λ

Radio wave travels at the speed of light (c)
$c = 3 \times 10^8$ meters/sec

$\lambda = c/f = 300 / f$ in MHz

Figure T3-2 — As a radio wave travels, it oscillates at the frequency of the signal. The distance covered by the wave during the time it takes for one complete cycle is its wavelength.

T3B02 **What term describes the number of times per second that an alternating current reverses direction?**

A. Pulse rate
B. Speed
C. Wavelength
D. Frequency

D One complete sequence of alternating current (ac) flowing in one direction, stopping, reversing, and stopping again is a *cycle*. The number of cycles per second is the ac current's frequency. This also describes the frequency of a radio signal or wave. [*Ham Radio License Manual*, page 3-6]

T3B03 **What are the two components of a radio wave?**

A. AC and DC
B. Voltage and current
C. Electric and magnetic fields
D. Ionizing and non-ionizing radiation

C See question T3A07. [*Ham Radio License Manual*, page 4-6]

T3B04 **How fast does a radio wave travel through free space?**

A. At the speed of light
B. At the speed of sound
C. Its speed is inversely proportional to its wavelength
D. Its speed increases as the frequency increases

A All electromagnetic energy — radio waves, light, X-rays — travels at the speed of light. In a vacuum, the speed of light (represented by a lower-case *c*) is 300,000,000 meters per second. It travels close to that speed in air. In denser materials such as water or glass, and in cables, light travels slower. [*Ham Radio License Manual*, page 2-4]

T3B05 **How does the wavelength of a radio wave relate to its frequency?**

A. The wavelength gets longer as the frequency increases
B. The wavelength gets shorter as the frequency increases
C. There is no relationship between wavelength and frequency
D. The wavelength depends on the bandwidth of the signal

B Because radio waves travel at a constant speed, the higher their frequency, the less time it takes to complete one cycle, and the less distance it has traveled during that cycle — its wavelength. Higher frequencies mean shorter wavelengths and vice versa. (See question T3A01.) [*Ham Radio License Manual*, page 2-5]

T3B06 **What is the formula for converting frequency to wavelength in meters?**

A. Wavelength in meters equals frequency in hertz multiplied by 300
B. Wavelength in meters equals frequency in hertz divided by 300
C. Wavelength in meters equals frequency in megahertz divided by 300
D. Wavelength in meters equals 300 divided by frequency in megahertz

D The key to remembering this relationship is that wavelength and frequency are inversely proportional. That is, as one gets larger, the other must get smaller. The formula that describes this relationship is: Wavelength (l, in meters) = 300 / (Frequency (f) in MHz). [*Ham Radio License Manual*, page 2-5]

T3B07 **What property of radio waves is often used to identify the different frequency bands?**

A. The approximate wavelength
B. The magnetic intensity of waves
C. The time it takes for waves to travel one mile
D. The voltage standing wave ratio of waves

A Because of the radio wave's constant velocity (see question T3A01), knowing a radio wave's wavelength is the same as knowing its frequency and vice versa. By referring to a frequency band by approximate wavelength, which is traditional, the frequency range is also implied. [*Ham Radio License Manual*, page 2-5]

T3B08 **What are the frequency limits of the VHF spectrum?**

A. 30 to 300 kHz
B. 30 to 300 MHz
C. 300 to 3000 kHz
D. 300 to 3000 MHz

B See the table of RF Spectrum Ranges. [*Ham Radio License Manual*, page 2-3]

RF Spectrum Ranges

Range Name	Abbreviation	Frequency Range
Very Low Frequency	VLF	3 kHz - 30 kHz
Low Frequency	LF	30 kHz - 300 kHz
Medium Frequency	MF	300 kHz - 3 MHz
High Frequency	HF	3 MHz - 30 MHz
Very High Frequency	VHF	30 MHz - 300 MHz
Ultra High Frequency	UHF	300 MHz - 3 GHz
Super High Frequency	SHF	3 GHz - 30 GHz
Extremely High Frequency	EHF	30 GHz - 300 GHz

T3B09   What are the frequency limits of the UHF spectrum?

A. 30 to 300 kHz
B. 30 to 300 MHz
C. 300 to 3000 kHz
D. 300 to 3000 MHz

D   See the table of RF Spectrum Ranges. [*Ham Radio License Manual*, page 2-3]

T3B10   What frequency range is referred to as HF?

A. 300 to 3000 MHz
B. 30 to 300 MHz
C. 3 to 30 MHz
D. 300 to 3000 kHz

C   See the table of RF Spectrum Ranges. [*Ham Radio License Manual*, page 2-3]

T3B11   What is the approximate velocity of a radio wave as it travels through free space?

A. 3000 kilometers per second
B. 300,000,000 meters per second
C. 300,000 miles per hour
D. 186,000 miles per hour

B   See question T3B04. [*Ham Radio License Manual*, page 2-4]

T3C   Propagation modes; line of sight, sporadic E, meteor, aurora scatter, tropospheric ducting, F layer skip, radio horizon

T3C01   Why are "direct" (not via a repeater) UHF signals rarely heard from stations outside your local coverage area?

A. They are too weak to go very far
B. FCC regulations prohibit them from going more than 50 miles
C. UHF signals are usually not reflected by the ionosphere
D. They collide with trees and shrubbery and fade out

C   The ability of the ionosphere to diffract or bend radio waves also depends on the frequency of the radio wave. Higher frequency waves are bent less than those of lower frequencies. At VHF and higher frequencies the waves usually pass through the ionosphere with only a little bending and are lost to space. This is why VHF and UHF signals from stations beyond the radio horizon are rarely heard without being relayed by a repeater. [*Ham Radio License Manual*, page 4-4]

T3C02   Which of the following might be happening when VHF signals are being received from long distances?

A. Signals are being reflected from outer space
B. Signals are arriving by sub-surface ducting
C. Signals are being reflected by lightning storms in your area
D. Signals are being refracted from a sporadic E layer

D   At all points in the solar cycle, patches of the ionosphere's E-layer can become sufficiently ionized to reflect VHF signals back to earth. Most common on the 6 meter and 2 meter bands, sporadic-E propagation occurs from reflections off highly ionized patches in the ionosphere's E-layer. The patches form at irregular intervals and places, earning them the name "sporadic." Sporadic-E propagation can relay signals as far as 1500 miles. [*Ham Radio License Manual*, page 4-4]

T3C03   What is a characteristic of VHF signals received via auroral reflection?

A. Signals from distances of 10,000 or more miles are common
B. The signals exhibit rapid fluctuations of strength and often sound distorted
C. These types of signals occur only during winter nighttime hours
D. These types of signals are generally strongest when your antenna is aimed to the south (for stations in the Northern Hemisphere

B   The aurora (northern lights) is the glow from thin sheets of charged particles flowing down through the lower layers of the ionosphere. Those thin sheets 50 miles or more above the Earth's surface reflect VHF and UHF signals. Because the aurora is constantly changing, the reflected signals change strength quickly and are often distorted. [*Ham Radio License Manual*, page 4-4]

T3C04   Which of the following propagation types is most commonly associated with occasional strong over-the-horizon signals on the 10, 6, and 2 meter bands?

A. Backscatter
B. Sporadic E
C. D layer absorption
D. Gray-line propagation

B   See question T3C02. [*Ham Radio License Manual*, page 4-4]

T3C05 **What is meant by the term "knife-edge" propagation?**

A. Signals are reflected back toward the originating station at acute angles
B. Signals are sliced into several discrete beams and arrive via different paths
C. Signals are partially refracted around solid objects exhibiting sharp edges
D. Signals propagated close to the band edge exhibiting a sharp cutoff

C Radio waves can be reflected by any sudden change in the media through which they are traveling, such as a building, hill or even weather-related changes in the atmosphere. Obstructions such as buildings and hills create radio shadows, especially at VHF and UHF frequencies. The figure shows how radio waves can also be diffracted or bent as they travel past sharp edges of these objects. This is called knife-edge propagation. [*Ham Radio License Manual*, page 4-1]

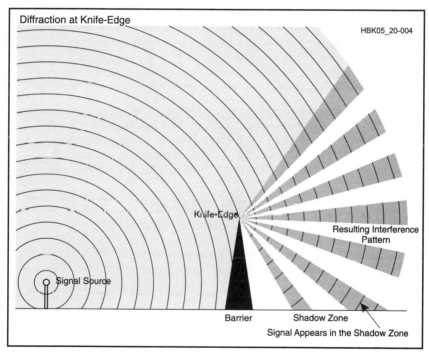

Diffraction at Knife-Edge

HBK05_20-004

Knife-Edge

Resulting Interference Pattern

Signal Source

Barrier   Shadow Zone

Signal Appears in the Shadow Zone

Figure T3-3 — VHF and UHF radio waves are diffracted around the sharp edge of a solid object, such as a building, hill or other obstruction. Some signals appear behind the obstruction as a result of interference between waves at the edge and those farther away. The resulting interference pattern creates shadowed areas where little signal is present.

T3C06   What mode is responsible for allowing over-the-horizon VHF and UHF communications to ranges of approximately 300 miles on a regular basis?

A. Tropospheric scatter
B. D layer refraction
C. F2 layer refraction
D. Faraday rotation

A   Propagation at and above VHF frequencies assisted by atmospheric phenomena such as weather fronts or temperature inversions is called tropospheric scatter propagation or just "tropo." Layers of air with different characteristics can also form structures called ducts that can guide even microwave signals for long distances. Tropo is regularly used by amateurs to make VHF and UHF contacts that would otherwise be impossible by line-of-sight propagation. Tropo contacts over 300-mile paths are not uncommon. [*Ham Radio License Manual*, page 4-2]

T3C07   What band is best suited to communicating via meteor scatter?

A. 10 meters
B. 6 meters
C. 2 meters
D. 70 cm

B   A meteoroid burning up in the upper atmosphere leaves a trail of ionized gas lasting up to several seconds that can reflect radio signals. Bouncing signals off of these ionized trails is called meteor scatter propagation. 6 meters is the best band for meteor scatter and contacts can be made at distances up to 1200 to 1500 miles. [*Ham Radio License Manual*, page 4-5]

T3C08   What causes "tropospheric ducting"?

A. Discharges of lightning during electrical storms
B. Sunspots and solar flares
C. Updrafts from hurricanes and tornadoes
D. Temperature inversions in the atmosphere

D   See question T3C06. [*Ham Radio License Manual*, page 4-2]

T3C09 **What is generally the best time for long-distance 10 meter band propagation?**

A. During daylight hours
B. During nighttime hours
C. When there are coronal mass ejections
D. Whenever the solar flux is low

A   The highest frequency signal that can be reflected back to a point on the earth is the maximum usable frequency (MUF) between the transmitter and receiver. The lowest frequency that can travel between those points without being absorbed is the lowest usable frequency (LUF). (See the figure below.) The MUF rises as the sun illuminates the ionosphere. That's why the upper HF bands, such as 10 meters, are more likely to be open during the day. [*Ham Radio License Manual*, page 4-4]

Figure T3-4 — Signals that are too low in frequency are absorbed by the ionosphere and lost. Signals that are too high in frequency pass through the ionosphere and are also lost. Signals in the right range of frequencies are refracted back toward the Earth and are received hundreds or thousands of miles away.

T3C10   What is the radio horizon?

   A. The distance at which radio signals between two points are effectively blocked by the curvature of the Earth
   B. The distance from the ground to a horizontally mounted antenna
   C. The farthest point you can see when standing at the base of your antenna tower
   D. The shortest distance between two points on the Earth's surface

A   Since most propagation at VHF and UHF frequencies is line-of-sight, the limit of the range of these signals is determined by the radio horizon. (There is some slight bending of radio waves along the Earth's surface, so the radio horizon is slightly more distant than the visual horizon.) Increasing the height of either the transmitting or receiving antennas also increases the radio horizon's distance. [*Ham Radio License Manual*, page 4-1]

T3C11   Why do VHF and UHF radio signals usually travel somewhat farther than the visual line of sight distance between two stations?

   A. Radio signals move somewhat faster than the speed of light
   B. Radio waves are not blocked by dust particles
   C. The Earth seems less curved to radio waves than to light
   D. Radio waves are blocked by dust particles

C   Diffraction makes the Earth seem less curved to VHF and UHF radio waves than light. This allows VHF and UHF signals to travel farther than the visual line-of-sight horizon. (See question T3C10.) [*Ham Radio License Manual*, page 4-1]

Station Setup and Operation

SUBELEMENT T4 — Amateur radio practices and station set up [2 exam questions — 2 groups]

T4A Station setup; microphone, speaker, headphones, filters, power source, connecting a computer, RF grounding

T4A01 Which of the following is true concerning the microphone connectors on amateur transceivers?

A. All transceivers use the same microphone connector type
B. Some connectors include push-to-talk and voltages for powering the microphone
C. All transceivers using the same connector type are wired identically
D. Un-keyed connectors allow any microphone to be connected

B The microphone connector of a transceiver is likely to include push-to-talk connections and also supply voltage for powering electret-style microphones. The wiring of the connector varies for the different manufacturers, so check the connections before trying a new microphone. [*Ham Radio License Manual*, page 5-4]

T4A02 What could be used in place of a regular speaker to help you copy signals in a noisy area?

A. A video display
B. A low pass filter
C. A set of headphones
D. A boom microphone

C Using headphones blocks wind noise and noise from nearby people or machinery. Remember that the same noise may be picked up by your microphone, so take steps to shield it from noise as well. [*Ham Radio License Manual*, page 2-13]

T4A03   Which is a good reason to use a regulated power supply for communications equipment?

A. It prevents voltage fluctuations from reaching sensitive circuits
B. A regulated power supply has FCC approval
C. A fuse or circuit breaker regulates the power
D. Power consumption is independent of load

A   To operate properly, communications equipment requires a power supply with a stable output voltage. Regulation prevents voltage fluctuations of the input power source from causing similar fluctuations at the output. [*Ham Radio License Manual*, page 5-14]

T4A04   Where must a filter be installed to reduce harmonic emissions?

A. Between the transmitter and the antenna
B. Between the receiver and the transmitter
C. At the station power supply
D. At the microphone

A   Due to minor imperfections, every transmitter's RF output signal contains weak harmonics of the desired output signal and other spurious emissions that can cause interference to nearby equipment. In extreme cases, a misadjusted or defective transmitter can generate strong interfering signals. To prevent harmonics from being radiated, a low-pass or band-pass filter must be installed at the transmitter's connection to the antenna feed line. [*Ham Radio License Manual*, page 5-20]

T4A05   What type of filter should be connected to a TV receiver as the first step in trying to prevent RF overload from a nearby 2 meter transmitter?

A. Low-pass filter
B. High-pass filter
C. Band-pass filter
D. Band-reject filter

D   If the interfering signal's frequency is close to the desired signal's frequency, it may not be possible to prevent overload with a high-pass or low-pass filter because the desired signal will be attenuated, as well. For example, when a TV receiver is overloaded by a nearby 2 meter transmitter, a notch or band-reject filter is required. The notch filter attenuates the 2 meter signal without also filtering out over-the-air broadcast channels 2 - 13. [*Ham Radio License Manual*, page 5-20]

T4A06 Which of the following would be connected between a transceiver and computer in a packet radio station?

A. Transmatch
B. Mixer
C. Terminal node controller
D. Antenna

C The terminal node controller (TNC) converts the audio signals from your radio's receiver output to data characters for the computer. Similarly, characters from the computer are converted to audio signals that your transmitter can use. The TNC may also act as the push-to-talk switch to turn the transmitter on and off. [*Ham Radio License Manual*, page 5-11]

T4A07 How is the computer's sound card used when conducting digital communications using a computer?

A. The sound card communicates between the computer CPU and the video display
B. The sound card records the audio frequency for video display
C. The sound card provides audio to the microphone input and converts received audio to digital form
D. All of these choices are correct

C Sound cards and their host computers have become sufficiently powerful that for several popular digital modes such as RTTY and PSK31, there is no need for a separate TNC or protocol controller. The sound card and software running on the host can perform the necessary conversions between audio signals and data characters. [*Ham Radio License Manual*, page 5-11]

T4A08 Which type of conductor is best to use for RF grounding?

A. Round stranded wire
B. Round copper-clad steel wire
C. Twisted-pair cable
D. Flat strap

D Connections to an RF ground bus should be made with a short, wide conductor such as copper flashing or strap or heavy solid wire (#8 AWG or larger). Solid strap is best because it presents the lowest impedance to RF current. [*Ham Radio License Manual*, page 5-24]

T4A09   Which would you use to reduce RF current flowing on the shield of an audio cable?

A. Band-pass filter
B. Low-pass filter
C. Preamplifier
D. Ferrite choke

D   RF choke or common-mode filters made of ferrite material are used to reduce RF currents flowing on unshielded wires such as speaker cables, ac power cords and telephone modular cords. Ferrite chokes are also used to reduce RF current flowing on the outside of shielded audio and computer cables. [*Ham Radio License Manual*, page 5-18]

T4A10   What is the source of a high-pitched whine that varies with engine speed in a mobile transceiver's receive audio?

A. The ignition system
B. The alternator
C. The electric fuel pump
D. Anti-lock braking system controllers

B   Alternator whine is a type of noise caused by noise on the dc power system inside your own vehicle. You might hear it with the received audio but more likely it will be heard by others as a high-pitched whine on your transmitted signal that varies with your engine speed. It can be removed by a dc power filter at your radio. [*Ham Radio License Manual*, page 5-15]

T4A11   Where should a mobile transceiver's power negative connection be made?

A. At the battery or engine block ground strap
B. At the antenna mount
C. To any metal part of the vehicle
D. Through the transceiver's mounting bracket

A   Connect the radio's negative lead to the negative battery terminal or where the battery ground lead is connected to the vehicle body. Do not rely on ground connections via the vehicle's body because vehicle sections may not be bonded together dependably. Many body components may be plastic or composite, as well. The most reliable negative power connection is directly to the battery or its ground strap. [*Ham Radio License Manual*, page 5-14]

T4B Operating controls; tuning, use of filters, squelch, AGC, repeater offset, memory channels

T4B01 What may happen if a transmitter is operated with the microphone gain set too high?

A. The output power might be too high
B. The signal might become distorted
C. The frequency might vary
D. The SWR might increase

B Your transmitter will produce spurious emissions if your microphone gain is too high. For example, on an SSB transmitter the resulting overmodulation creates spurious signals on nearby frequencies. On FM, the result will be excessive deviation that can cause your signal to encroach on an adjacent channel. [*Ham Radio License Manual*, page 5-4]

T4B02 Which of the following can be used to enter the operating frequency on a modern transceiver?

A. The keypad or VFO knob
B. The CTCSS or DTMF encoder
C. The Automatic Frequency Control
D. All of these choices are correct

A The VFO (variable frequency oscillator) is the circuit that controls the frequency of operation on both receive and transmit. Many radios also have a numeric keypad that allows you to enter the desired frequency directly. The VFO is used when you are tuning across the band looking for a station or particular signal. Keypads are used when changing directly between known frequencies. [*Ham Radio License Manual*, page 5-2]

T4B03 What is the purpose of the squelch control on a transceiver?

A. To set the highest level of volume desired
B. To set the transmitter power level
C. To adjust the automatic gain control
D. To quiet the noise when no signal is being received

D In the absence of a signal on an FM receiver, you will hear noise. FM receivers have a squelch circuit that cuts off the speaker unless a signal is present, called "closing" the squelch. The squelch control adjusts the threshold at which squelch circuits turn the speaker on and off. The proper setting for the squelch control is just above the point at which the receiver audio is cut off. If set at a higher level, some signals will not be strong enough "open" the squelch. [*Ham Radio License Manual*, page 5-6]

T4B04   What is a way to enable quick access to a favorite frequency on your transceiver?

    A. Enable the CTCSS tones
    B. Store the frequency in a memory channel
    C. Disable the CTCSS tones
    D. Use the scan mode to select the desired frequency

B   By storing the frequency and any other pertinent information in a memory channel you can quickly return to the frequency at any time. [*Ham Radio License Manual*, page 5-2]

T4B05   Which of the following would reduce ignition interference to a receiver?

    A. Change frequency slightly
    B. Decrease the squelch setting
    C. Turn on the noise blanker
    D. Use the RIT control

C   Ignition noise consists of a sharp pulse each time a spark plug in the engine fires. Just like the early spark radio transmitters, the arc produces RF signals across a wide range of frequencies. You hear the signals as buzzing or raspy noise that changes pitch along with engine speed. The noise blanker circuit in your radio detects these pulses and mutes the receiver for a short period. [*Ham Radio License Manual*, page 5-7]

T4B06   Which of the following controls could be used if the voice pitch of a single-sideband signal seems too high or low?

    A. The AGC or limiter
    B. The bandwidth selection
    C. The tone squelch
    D. The receiver RIT or clarifier

D   Receiver incremental tuning (RIT) is a fine-tuning control used for SSB or CW operation. RIT allows the operator to adjust the receiver frequency without changing the transmitter frequency. This allows you to tune in a station that is slightly off frequency or to adjust the pitch of an operator's voice that seems too high or low. [*Ham Radio License Manual*, page 5-6]

T4B07   What does the term "RIT" mean?

    A. Receiver Input Tone
    B. Receiver Incremental Tuning
    C. Rectifier Inverter Test
    D. Remote Input Transmitter

B   See question T4B06. [*Ham Radio License Manual*, page 5-6]

T4B08   What is the advantage of having multiple receive bandwidth choices on a multimode transceiver?

A. Permits monitoring several modes at once
B. Permits noise or interference reduction by selecting a bandwidth matching the mode
C. Increases the number of frequencies that can be stored in memory
D. Increases the amount of offset between receive and transmit frequencies

B   IF filters are specified by bandwidth in Hz or kHz. Wide filters (around 2.4 kHz) are used for SSB reception on phone. Narrow filters (around 500 Hz) are used for Morse code and data mode reception. Having multiple filters available allows you to reduce noise or interference by selecting a filter with just enough bandwidth to pass the desired signal. [*Ham Radio License Manual*, page 5-6]

T4B09   Which of the following is an appropriate receive filter to select in order to minimize noise and interference for SSB reception?

A. 500 Hz
B. 1000 Hz
C. 2400 Hz
D. 5000 Hz

C   See question T4B08. [*Ham Radio License Manual*, page 5-6]

T4B10   Which of the following is an appropriate receive filter to select in order to minimize noise and interference for CW reception?

A. 500 Hz
B. 1000 Hz
C. 2400 Hz
D. 5000 Hz

A   See question T4B08. [*Ham Radio License Manual*, page 5-6]

T4B11   Which of the following describes the common meaning of the term "repeater offset"?

A. The distance between the repeater's transmit and receive antennas
B. The time delay before the repeater timer resets
C. The difference between the repeater's transmit and receive frequencies
D. The maximum frequency deviation permitted on the repeater's input signal

C   To send a signal through a repeater, you must transmit on the repeater's input frequency where the repeater receiver listens, called the repeater's input frequency. You then listen where the repeater transmits, called the repeater's output frequency. To make it easy to use many different repeaters, hams have decided on a standard separation between input and output frequencies. The difference between repeater input and output frequencies is called the repeater's offset or shift. (See question T2A01.) [*Ham Radio License Manual*, page 6-16]

Electrical and Electronic Principles

SUBELEMENT T5 — Electrical principles, math for electronics, electronic principles, Ohm's Law [4 exam questions, 4 groups]

T5A Electrical principles; current and voltage, conductors and insulators, alternating and direct current

T5A01 Electrical current is measured in which of the following units?

    A. Volts
    B. Watts
    C. Ohms
    D. Amperes

D   The basic unit of electric current, a measure of the rate of flow of electrons, is the ampere, abbreviated A. (Current is represented by the symbol *I* or *i* in equations.) It is named for Andre Marie Ampere, an early 19th century scientist who studied electricity extensively. [*Ham Radio License Manual*, page 3-1]

T5A02 Electrical power is measured in which of the following units?

    A. Volts
    B. Watts
    C. Ohms
    D. Amperes

B   Power is the rate at which energy is used. The basic unit of electrical power is the watt, or W. (Power is represented by the symbol *p* in equations.) This unit was named after James Watt, the 18th century inventor of the steam engine. [*Ham Radio License Manual*, page 3-5]

T5A03 **What is the name for the flow of electrons in an electric circuit?**

A. Voltage
B. Resistance
C. Capacitance
D. Current

D Electrons flow through the wires and components of an electric circuit. The flow of electrons in an electric circuit is called *current*. (See question T5A01.) [*Ham Radio License Manual*, page 3-1]

T5A04 **What is the name for a current that flows only in one direction?**

A. Alternating current
B. Direct current
C. Normal current
D. Smooth current

B Current that flows in one direction all the time is *direct current*, abbreviated *dc*. [*Ham Radio License Manual*, page 3-5]

T5A05 **What is the electrical term for the electromotive force (EMF) that causes electron flow?**

A. Voltage
B. Ampere-hours
C. Capacitance
D. Inductance

A *Voltage* is the electromotive force or electric potential that makes electrons move. Voltage is measured in volts, abbreviated as V. Voltage is represented in equations by the symbol *E* or *e* or *v*. [*Ham Radio License Manual*, page 3-1]

T5A06 **How much voltage does a mobile transceiver usually require?**

A. About 12 volts
B. About 30 volts
C. About 120 volts
D. About 240 volts

A Vehicle power systems are usually referred to as "12 volt", although the actual voltage typically varies from 12 to 15 V. Radios designed to operate from a "12 V" supply may actually work best at the slightly higher voltage of 13.8 V, typical of vehicle power systems with the engine running. Check the equipment manual and be sure your power supply can generate the correct voltage. [*Ham Radio License Manual*, page 5-14]

T5A07   Which of the following is a good electrical conductor?

A. Glass
B. Wood
C. Copper
D. Rubber

C In general, most metals make good conductors of electricity because their electrons are relatively free to move in response to an applied voltage. Copper, silver and aluminum are all excellent conductors, while wood, paper and glass are very poor conductors. [*Ham Radio License Manual*, page 3-4]

T5A08   Which of the following is a good electrical insulator?

A. Copper
B. Glass
C. Aluminum
D. Mercury

B Most non-metallic materials act as insulators because their electrons are not free to move. That means very little current will flow in response to an applied voltage. Ceramics, glass, paper and plastics are examples of good insulators. [*Ham Radio License Manual*, page 3-4]

T5A09   What is the name for a current that reverses direction on a regular basis?

A. Alternating current
B. Direct current
C. Circular current
D. Vertical current

A *Alternating current*, or *ac*, alternates direction, flowing first in one direction, then in the opposite direction. [*Ham Radio License Manual*, page 3-5]

T5A10   Which term describe the rate at which electrical energy is used?

A. Resistance
B. Current
C. Power
D. Voltage

C See question T5A02. [*Ham Radio License Manual*, page 3-5]

T5A11 What is the basic unit of electromotive force?

A. The volt
B. The watt
C. The ampere
D. The ohm

A   See question T5A05. [*Ham Radio License Manual*, page 3-1]

T5B   Math for electronics; decibels, electrical units and the metric system

T5B01 How many milliamperes is 1.5 amperes?

A. 15 milliamperes
B. 150 milliamperes
C. 1,500 milliamperes
D. 15,000 milliamperes

C   The prefix "milli" means "divide by 1000," so "milliamperes" means "divide by 1000 to get amperes." 1500 milliamperes / 1000 = 1.5 amperes. Similarly, to convert from amperes to milliamperes, multiply by 1000, so 1.5 amperes × 1000 = 1500 milliamperes. [*Ham Radio License Manual*, page 2-2]

International System of Units (SI) — Metric Units

Prefix	Symbol	Multiplication Factor		
Tera	T	10^{12}	=	1,000,000,000,000
Giga	G	10^{9}	=	1,000,000,000
Mega	M	10^{6}	=	1,000,000
Kilo	k	10^{3}	=	1000
Hecto	h	10^{2}	=	100
Deca	da	10^{1}	=	10
Deci	d	10^{-1}	=	0.1
Centi	c	10^{-2}	=	0.01
Milli	m	10^{-3}	=	0.001
Micro	μ	10^{-6}	=	0.000001
Nano	n	10^{-9}	=	0.000000001
Pico	p	10^{-12}	=	0.000000000001

T5B02　**What is another way to specify a radio signal frequency of 1,500,000 hertz?**

A. 1500 kHz
B. 1500 MHz
C. 15 GHz
D. 150 kHz

A　Checking each of the possible answers:

1500 kHz × 1000 = 1,500,000 Hz (the correct answer)

1500 MHz × 1,000,000 = 1,500,000,000 Hz (too large, 1,500,000 Hz = 1.5 MHz)

15 GHz × 1,000,000,000 = 1,500,000,000,000 Hz (very much too large, 1,500,000 Hz = 0.0015 GHz)

150 kHz × 1000 = 150,000 Hz (too small)

[*Ham Radio License Manual*, page 2-2]

T5B03　**How many volts are equal to one kilovolt?**

A. One one-thousandth of a volt
B. One hundred volts
C. One thousand volts
D. One million volts

C　The prefix "kilo" means "multiply by 1000," so "kilovolts" means "multiply by 1000 to get volts." 1 kilovolt × 1000 = 1000 volts. [*Ham Radio License Manual*, page 2-2]

T5B04　**How many volts are equal to one microvolt?**

A. One one millionth of a volt
B. One million volts
C. One thousand kilovolts
D. One one-thousandth of a volt

A　The prefix "micro" means "divide by 1,000,000," so "microvolts" means "divide by 1,000,000 to get volts." 1 microvolt / 1,000,000 = 0.000001 V or one-millionth of a volt. [*Ham Radio License Manual*, page 2-2]

T5B05　**Which of the following is equivalent to 500 milliwatts?**

A. 0.02 watts
B. 0.5 watts
C. 5 watts
D. 50 watts

B　The prefix "milli" means "divide by 1000," so "milliwatts" means "divide by 1000 to get watts." 500 milliwatts / 1000 = 0.5 W or watts. [*Ham Radio License Manual*, page 2-2]

T5B06 If an ammeter calibrated in amperes is used to measure a 3000-milliampere current, what reading would it show?

A. 0.003 amperes
B. 0.3 amperes
C. 3 amperes
D. 3,000,000 amperes

C The prefix "milli" means "divide by 1000," so "milliwatts" means "divide by 1000 to get watts." 3000 milliamperes / 1000 = 3 A or amperes. [*Ham Radio License Manual*, page 2-2]

T5B07 If a frequency readout calibrated in megahertz shows a reading of 3.525 MHz, what would it show if it were calibrated in kilohertz?

A. 0.003525 kHz
B. 35.25 kHz
C. 3525 kHz
D. 3,525,000 kHz

C To convert from MHz to kHz, multiply by 1000 because 1000 kHz = 1 MHz. 3.525 × 1000 = 3525 kHz. [*Ham Radio License Manual*, page 2-2]

T5B08 How many microfarads are 1,000,000 picofarads?

A. 0.001 microfarads
B. 1 microfarad
C. 1000 microfarads
D. 1,000,000,000 microfarads

B To convert from picofarads to microfarads divide by 1,000,000, so 1,000,000 picofarads (or pF) = 1 microfarad (or μF). [*Ham Radio License Manual*, page 2-2]

T5B09 What is the approximate amount of change, measured in decibels (dB), of a power increase from 5 watts to 10 watts?

A. 2 dB
B. 3 dB
C. 5 dB
D. 10 dB

B Radio signals vary dramatically in strength. At the input to a receiver, signals are frequently smaller than one ten-billionth of a watt. When they come out of a transmitter, they're often measured in kilowatts! Electronic circuits change signal strengths by many factors of 10. These big differences in value make it difficult to compare signal sizes. Enter the decibel, abbreviated dB and pronounced "dee-bee." The decibel measures the ratio of two quantities as a power of 10. The formula for computing decibels is:

$dB = 10 \log (\text{power ratio})$

$dB = 20 \log (\text{voltage ratio})$.

Positive values of dB mean the ratio is greater than 1 and negative values of dB indicate a ratio of less than 1. For example, if an amplifier turns a 5-watt signal into a 10-watt signal, that's a change of $10 \log (10 / 5) = 10 \log (2) = 3$ dB. [*Ham Radio License Manual*, page 4-6]

T5B10 What is the approximate amount of change, measured in decibels (dB), of a power decrease from 12 watts to 3 watts?

A. 1 dB
B. 3 dB
C. 6 dB
D. 9 dB

C $dB = 10 \log (3 / 12) = 10 \log (0.25) = -9$ dB, so the amount of change is 9 dB (regardless of whether the amount of change is positive or negative). [*Ham Radio License Manual*, page 4-6]

T5B11 What is the approximate amount of change, measured in decibels (dB), of a power increase from 20 watts to 200 watts?

A. 10 dB
B. 12 dB
C. 18 dB
D. 28 dB

A $dB = 10 \log (200 / 20) = 10 \log (10) = 10$ dB [*Ham Radio License Manual*, page 4-6]

T5C   Electronic principles; capacitance, inductance, current flow in circuits, alternating current, definition of RF, power calculations

T5C01   What is the ability to store energy in an electric field called?

A. Inductance
B. Resistance
C. Tolerance
D. Capacitance

D   The ability to store energy in an electric field is called capacitance and it is measured in farads (F). *Capacitors* store electrical energy in the electric field created by a voltage between two conducting surfaces (such as metal foil) called *electrodes*. The electrodes are separated by an insulator called the *dielectric*. [*Ham Radio License Manual*, page 3-6]

T5C02   What is the basic unit of capacitance?

A. The farad
B. The ohm
C. The volt
D. The henry

A   See question T5C01. [*Ham Radio License Manual*, page 3-6]

T5C03   What is the ability to store energy in a magnetic field called?

A. Admittance
B. Capacitance
C. Resistance
D. Inductance

D   *Inductors* store magnetic energy in the magnetic field created by current flowing through a wire. This property is called *inductance* and it is measured in henrys (H). [*Ham Radio License Manual*, page 3-7]

T5C04   What is the basic unit of inductance?

A. The coulomb
B. The farad
C. The henry
D. The ohm

C   See question T5C03. [*Ham Radio License Manual*, page 3-7]

T5C05   What is the unit of frequency?

A. Hertz
B. Henry
C. Farad
D. Tesla

A   Each complete back-and-forth sequence of a radio signal (or any oscillating system, such as a pendulum) is called a cycle. The number of cycles per second is the signal's *frequency*, represented by a lower-case *f*. The unit of measurement for frequency is the *hertz*, abbreviated Hz. One cycle per second is one hertz or 1 Hz. As frequency increases, it becomes easier to use units of kilohertz (1 kHz = 1000 Hz), megahertz (1 MHz = 1000 kHz = 1,000,000 Hz), and gigahertz (1 GHz = 1000 MHz = 1,000,000,000 Hz). [*Ham Radio License Manual*, page 2-2]

T5C06   What is the abbreviation that refers to radio frequency signals of all types?

A. AF
B. HF
C. RF
D. VHF

C   If connected to a speaker, signals below 20 kHz produce sound waves you can hear, so we call them *audio frequency* or *AF* signals. Signals that have a frequency greater than 20,000 Hz (or 20 kHz) are *radio frequency* or *RF* signals. [*Ham Radio License Manual*, page 2-3]

T5C07   What is a usual name for electromagnetic waves that travel through space?

A. Gravity waves
B. Sound waves
C. Radio waves
D. Pressure waves

C   A radio wave is a combination of an electric and a magnetic field, just like the fields in a capacitor and inductor but spreading out into space like ripples traveling across the surface of water. Because a radio wave is made up of both types of fields, it is called an electromagnetic wave. (See questions T3A07 and T3B03.) [*Ham Radio License Manual*, page 3-5]

T5C08 **What is the formula used to calculate electrical power in a DC circuit?**

A. Power (P) equals voltage (E) multiplied by current (I)
B. Power (P) equals voltage (E) divided by current (I)
C. Power (P) equals voltage (E) minus current (I)
D. Power (P) equals voltage (E) plus current (I)

A Power is the product of voltage and current. As with Ohm's Law, if you know any two of P, E or I, you can determine the missing quantity as follows: P = E × I or E = P / I or I = P / E. Voltage and current are never combined by adding or subtraction. [*Ham Radio License Manual*, page 3-5]

ARRL0496

Figure T5-1 -- This simple diagram will help you remember the power relationships. If you know any two of the quantities, you can find the third by covering up the unknown quantity. The positions of the remaining two symbols show if you have to multiply (side-by-side) or divide (one over the other).

T5C09 **How much power is being used in a circuit when the applied voltage is 13.8 volts DC and the current is 10 amperes?**

A. 138 watts
B. 0.7 watts
C. 23.8 watts
D. 3.8 watts

A Power = voltage × current = 13.8 V × 10 A = 138 W. [*Ham Radio License Manual*, page 3-5]

T5C10 How much power is being used in a circuit when the applied voltage is 12 volts DC and the current is 2.5 amperes?

A. 4.8 watts
B. 30 watts
C. 14.5 watts
D. 0.208 watts

B Power = voltage × current = 12 V × 2.5 A = 30 W. [*Ham Radio License Manual*, page 3-5]

T5C11 How many amperes are flowing in a circuit when the applied voltage is 12 volts DC and the load is 120 watts?

A. 0.1 amperes
B. 10 amperes
C. 12 amperes
D. 132 amperes

B From the equations presented in the explanation of question T5C08, current = power / voltage = 120 W / 12 V = 10 A. [*Ham Radio License Manual*, page 3-5]

T5D Ohm's Law

T5D01 What formula is used to calculate current in a circuit?

A. Current (I) equals voltage (E) multiplied by resistance (R)
B. Current (I) equals voltage (E) divided by resistance (R)
C. Current (I) equals voltage (E) added to resistance (R)
D. Current (I) equals voltage (E) minus resistance (R)

B Ohm's Law states that R = E / I. If you know any two of I, E, or R, you can determine the missing quantity as follows: R = E / I, I = E / R, or E = I × R. The drawing in the figure below is a convenient aid to remembering Ohm's Law in any of these three forms. [*Ham Radio License Manual*, page 3-4]

Figure T5-2 — This simple diagram will help you remember the Ohm's Law relationships. If you know any two of the quantities, you can find the third by covering up the unknown quantity. The positions of the remaining two symbols show if you have to multiply (side-by-side) or divide (one above the other).

T5D02   What formula is used to calculate voltage in a circuit?

A. Voltage (E) equals current (I) multiplied by resistance (R)
B. Voltage (E) equals current (I) divided by resistance (R)
C. Voltage (E) equals current (I) added to resistance (R)
D. Voltage (E) equals current (I) minus resistance (R)

A   Voltage = current × resistance, or E = I × R. (See question T5D01.) [*Ham Radio License Manual*, page 3-4]

T5D03   What formula is used to calculate resistance in a circuit?

A. Resistance (R) equals voltage (E) multiplied by current (I)
B. Resistance (R) equals voltage (E) divided by current (I)
C. Resistance (R) equals voltage (E) added to current (I)
D. Resistance (R) equals voltage (E) minus current (I)

B   Resistance = voltage divided by current, or R = E / I. (See question T5D01.) [*Ham Radio License Manual*, page 3-4]

T5D04   What is the resistance of a circuit in which a current of 3 amperes flows through a resistor connected to 90 volts?

A. 3 ohms
B. 30 ohms
C. 93 ohms
D. 270 ohms

B   R = E / I = 90 V / 3 A = 30 Ω or ohms. (See question T5D01.) [*Ham Radio License Manual*, page 3-4]

T5D05   What is the resistance in a circuit for which the applied voltage is 12 volts and the current flow is 1.5 amperes?

A. 18 ohms
B. 0.125 ohms
C. 8 ohms
D. 13.5 ohms

C   R = E / I = 12 V / 1.5 A = 8 Ω or ohms. (See question T5D01.) [*Ham Radio License Manual*, page 3-4]

T5D06   What is the resistance of a circuit that draws 4 amperes from a 12-volt source?

A. 3 ohms
B. 16 ohms
C. 48 ohms
D. 8 ohms

A   R = E / I = 12 V / 4 A = 3 Ω or ohms (See question T5D01.) [*Ham Radio License Manual*, page 3-4]

T5D07 **What is the current flow in a circuit with an applied voltage of 120 volts and a resistance of 80 ohms?**

A. 9600 amperes
B. 200 amperes
C. 0.667 amperes
D. 1.5 amperes

D $I = E / R = 120$ V $/ 80$ $\Omega = 1.5$ A or amperes. (See question T5D01.) [*Ham Radio License Manual*, page 3-4]

T5D08 **What is the current flowing through a 100-ohm resistor connected across 200 volts?**

A. 20,000 amperes
B. 0.5 amperes
C. 2 amperes
D. 100 amperes

C $I = E / R = 200$ V $/ 100$ $\Omega = 2$ A or amperes. (See question T5D01.) [*Ham Radio License Manual*, page 3-4]

T5D09 **What is the current flowing through a 24-ohm resistor connected across 240 volts?**

A. 24,000 amperes
B. 0.1 amperes
C. 10 amperes
D. 216 amperes

C $I = E / R = 240$ V $/ 24$ $\Omega = 10$ A or amperes. (See question T5D01) [*Ham Radio License Manual*, page 3-4]

T5D10 **What is the voltage across a 2-ohm resistor if a current of 0.5 amperes flows through it?**

A. 1 volt
B. 0.25 volts
C. 2.5 volts
D. 1.5 volts

A $E = I \times R = 0.5$ A $\times 2$ $\Omega = 1$ V or volt. (See question T5D01.) [*Ham Radio License Manual*, page 3-4]

T5D11   **What is the voltage across a 10-ohm resistor if a current of 1 ampere flows through it?**

A. 1 volt
B. 10 volts
C. 11 volts
D. 9 volts

B $E = I \times R = 1\,A \times 10\,\Omega = 10\,V$ or volts. (See question T5D01.) [*Ham Radio License Manual*, page 3-4]

T5D12   **What is the voltage across a 10-ohm resistor if a current of 2 amperes flows through it?**

A. 8 volts
B. 0.2 volts
C. 12 volts
D. 20 volts

D $E = I \times R = 2\,A \times 10\,\Omega = 20\,V$ or volts. (See question T5D01.) [*Ham Radio License Manual*, page 3-4]

Electrical Components and Functions

SUBELEMENT T6 — Electrical components, semiconductors, circuit diagrams, component functions
[4 Exam Questions — 4 Groups]

T6A Electrical components; fixed and variable resistors, capacitors, and inductors; fuses, switches, batteries

T6A01 What electrical component is used to oppose the flow of current in a DC circuit?

A. Inductor
B. Resistor
C. Voltmeter
D. Transformer

B *Resistors* have a certain resistance specified in ohms (Ω), kilo-ohms (kΩ), or mega-ohms (MΩ). The function of a resistor is to oppose the flow of electrical current, just as a valve in a water pipe restricts the flow through the pipe. [*Ham Radio License Manual*, page 3-6]

T6A02 What type of component is often used as an adjustable volume control?

A. Fixed resistor
B. Power resistor
C. Potentiometer
D. Transformer

C A variable resistor is also called a *potentiometer* (poh-ten-chee-AH-meh-tur) or *pot* because it is frequently used to adjust voltage or potential, such as for a volume control. [*Ham Radio License Manual*, page 3-8]

T6A03   What electrical parameter is controlled by a potentiometer?

A. Inductance
B. Resistance
C. Capacitance
D. Field strength

B   See question T6A02. [*Ham Radio License Manual*, page 3-8]

T6A04   What electrical component stores energy in an electric field?

A. Resistor
B. Capacitor
C. Inductor
D. Diode

B   *Capacitors* store electrical energy in the *electric field* created by a voltage between two conducting surfaces (such as metal foil) called *electrodes*. The electrodes are separated by an insulator called the *dielectric*. (See question T5C01.) [*Ham Radio License Manual*, page 3-6]

T6A05   What type of electrical component consists of two or more conductive surfaces separated by an insulator?

A. Resistor
B. Potentiometer
C. Oscillator
D. Capacitor

D   See question T6A04. [*Ham Radio License Manual*, page 3-6]

T6A06   What type of electrical component stores energy in a magnetic field?

A. Resistor
B. Capacitor
C. Inductor
D. Diode

C   *Inductors* store magnetic energy in the *magnetic field* created by current flowing through a wire. This is called *inductance* and it is measured in henrys (H). Inductors are made from wire wound in a coil, sometimes around a core of magnetic material that concentrates the magnetic energy. (See question T5C03.) [*Ham Radio License Manual*, page 3-7]

T6A07   What electrical component is usually composed of a coil of wire?

A. Switch
B. Capacitor
C. Diode
D. Inductor

D   See question T6A06. [*Ham Radio License Manual*, page 3-7]

T6A08   What electrical component is used to connect or disconnect electrical circuits?

A. Zener diode
B. Switch
C. Inductor
D. Variable resistor

B   *Switches* and *relays* are simple components that control current through a circuit by connecting and disconnecting the paths current can follow. Both can interrupt current — called *opening* a circuit — or allow it to flow — called *closing* a circuit. [*Ham Radio License Manual*, page 3-11]

T6A09   What electrical component is used to protect other circuit components from current overloads?

A. Fuse
B. Capacitor
C. Shield
D. Inductor

A   *Fuses* interrupt current overloads by melting a short length of metal. When the metal melts or "blows," the current path is broken and power is removed from circuits supplied by the fuse. [*Ham Radio License Manual*, page 3-11]

T6A10   What is the nominal voltage of a fully charged nickel-cadmium cell?

A. 1.0 volts
B. 1.2 volts
C. 1.5 volts
D. 2.2 volts

B   As seen in the following table, nickel-cadmium (NiCd) cells have a nominal voltage of 1.2 V when fully charged. [*Ham Radio License Manual*, page 5-17]

Battery Types and Characteristics

Battery Style	Chemistry Type	Fully-Charged Voltage	Energy Rating (average)
AAA	Alkaline — Disposable	1.5 V	1100 mAh
AA	Alkaline — Disposable	1.5 V	2600-3200 mAh
AA	Carbon-Zinc — Disposable	1.5 V	600 mAh
AA	Nickel-Cadmium (NiCd) — Rechargeable	1.2 V	700 mAh
AA	Nickel-Metal Hydride (NiMH) — Rechargeable	1.2 V	1500-2200 mAh
C	Alkaline — Disposable	1.5 V	7500 mAh
D	Alkaline — Disposable	1.5 V	14000 mAh
9 V	Alkaline — Disposable	9 V	580 mAh
9 V	Nickel-Cadmium (NiCd) — Rechargeable	9 V	110 mAh
9 V	Nickel-Metal Hydride — Rechargeable	9 V	150 mAh
Coin Cells	Lithium — Disposable	3-3.3 V	25-1000 mA
Packs	Lithium Ion — Rechargeable	3.3-3.6 V per cell	

T6A11 Which battery type is not rechargeable?

A. Nickel-cadmium
B. Carbon-zinc
C. Lead-acid
D. Lithium-ion

B Carbon-zinc batteries are disposable, not rechargeable, as shown in the table of battery types and characteristics. [*Ham Radio License Manual*, page 5-17]

T6B Semiconductors; basic principles of diodes and transistors

T6B01 What class of electronic components is capable of using a voltage or current signal to control current flow?

A. Capacitors
B. Inductors
C. Resistors
D. Transistors

D *Transistors* are components made from patterns of N- and P-type semiconductor material creating by *doping* — the adding of impurities to change the way the material conducts electricity. The patterns form structures that allow the transistor to use small voltages and currents to control larger ones. The transistor's electrodes are contacts made to a separate piece of the pattern. With the appropriate external circuit and a source of power, transistors can amplify or switch voltages and currents. Using small signals to control larger signals is called *gain*. [*Ham Radio License Manual*, page 3-10]

T6B02 **What electronic component allows current to flow in only one direction?**

A. Resistor
B. Fuse
C. Diode
D. Driven Element

C When N- and P-type material are placed in contact with each other, the result is a *PN junction* that conducts better in one direction than the other. This and other properties are used to create many useful electronic components, generally referred to as *semiconductors*. A semiconductor that only allows current flow in one direction is called a *diode*. Current flows from the *anode* to the *cathode*. [*Ham Radio License Manual*, page 3-10]

T6B03 **Which of these components can be used as an electronic switch or amplifier?**

A. Oscillator
B. Potentiometer
C. Transistor
D. Voltmeter

C See question T6B01. [*Ham Radio License Manual*, page 3-10]

T6B04 **Which of these components is made of three layers of semiconductor material?**

A. Alternator
B. Bipolar junction transistor
C. Triode
D. Pentagrid converter

B There are two common types of transistors: *bipolar junction transistors* (BJT) and field-effect transistors (FET). The BJT is made from three alternating layers of N- and P-type material. The electrodes are called the *base, emitter* and *collector*. There are two types of BJT; NPN and PNP, referring to the arrangement of the layers. [*Ham Radio License Manual*, page 3-10]

T6B05 **Which of the following electronic components can amplify signals?**

A. Transistor
B. Variable resistor
C. Electrolytic capacitor
D. Multi-cell battery

A See question T6B01. [*Ham Radio License Manual*, page 3-10]

T6B06 **How is a semiconductor diode's cathode lead usually identified?**

A. With the word "cathode"
B. With a stripe
C. With the letter "C"
D. All of these choices are correct

B Most semiconductor diode components are made with a protective coating around the actual semiconductor material. The electrodes are a pair of wire leads or tabs that make contact to the semiconductor material. Short of actually performing an electrical test, there is no way to tell the electrodes apart except for marking or labeling. The most common method is marking the body of the diode with a paint stripe to indicate the cathode electrode. [*Ham Radio License Manual*, page 3-10]

T6B07 **What does the abbreviation "LED" stand for?**

A. Low Emission Diode
B. Light Emitting Diode
C. Liquid Emission Detector
D. Long Echo Delay

B The *light-emitting diode* or LED, gives off light when current flows through it. [*Ham Radio License Manual*, page 3-10]

T6B08 **What does the abbreviation "FET" stand for?**

A. Field Effect Transistor
B. Fast Electron Transistor
C. Free Electron Transition
D. Field Emission Thickness

A The FET (or *field effect transistor*) is constructed as a conducting path or channel of N- or P-type material. The ends of the channel form the *source* and *drain* electrodes. The *gate* electrode is used to control current flow through the channel. [*Ham Radio License Manual*, page 3-10]

T6B09 **What are the names of the two electrodes of a diode?**

A. Plus and minus
B. Source and drain
C. Anode and cathode
D. Gate and base

C See question T6B02. [*Ham Radio License Manual*, page 3-10]

T6B10 Which semiconductor component has an emitter electrode?

A. Bipolar transistor
B. Field effect transistor
C. Silicon diode
D. Bridge rectifier

A   See question T6B04. [*Ham Radio License Manual*, page 3-10]

T6B11 Which semiconductor component has a gate electrode?

A. Bipolar transistor
B. Field effect transistor
C. Silicon diode
D. Bridge rectifier

B   See question T6B08. [*Ham Radio License Manual*, page 3-10]

T6B12 What is the term that describes a transistor's ability to amplify a signal?

A. Gain
B. Forward resistance
C. Forward voltage drop
D. On resistance

A   See question T6B01. [*Ham Radio License Manual*, page 3-10]

T6C   Circuit diagrams; schematic symbols

T6C01 What is the name for standardized representations of components in an electrical wiring diagram?

A. Electrical depictions
B. Grey sketch
C. Schematic symbols
D. Component callouts

C   If a circuit contains more than two or three components, trying to describe it clearly in words becomes very difficult. To describe complicated circuits, engineers have developed the *schematic diagram* or simply *schematic*. Schematics create a visual description of a circuit by using standardized representations of electrical components called circuit symbols. Figure T6-1 shows commonly used schematic symbols. [*Ham Radio License Manual*, page 3-12]

Schematic Symbols Used in Circuit Diagrams

Labeling conventions:
is a sequential number. **(X#)** is the component designator. Examples - C3, L11, R8, Q3

RESISTORS (R#)
FIXED
POTENTIOMETER
ADJUSTABLE
VARISTOR

CAPACITORS (C#)
FIXED
NON-POLARIZED
ELECTROLYTIC
VARIABLE

RELAYS (K#)
SPST
SPDT

GROUNDS
CHASSIS
EARTH
COMMON

SWITCHES (S#)
SPST
SPDT
DPST
DPDT

INDUCTORS (L#)
AIR-CORE
ADJUSTABLE
MAGNETIC-CORE
OR
FERRITE-BEAD (FB#)
TRANSFORMER (T#)

BATTERIES (BT#)
SINGLE CELL
MULTI CELL

* =
A - ANALOG
D - DIGITAL

DIODES (D#)
LED (DS#)
DIODE/RECTIFIER
ZENER
SCHOTTKY
(U#) or (D#)
BRIDGE RECTIFIERS
(U#)

TRANSISTORS (Q#)

B - Base,   C - Collector,   E - Emitter
G - Gate,   D - Drain,   S - Source

PNP
NPN
BIPOLAR JUNCTION

P-CHANNEL
N-CHANNEL
JUNCTION FET

P-CHANNEL
N-CHANNEL
SINGLE-GATE ENHANCEMENT MODE MOSFET

120 V AC WIRING
FEMALE (J#)
Black
HOT
120 V
NEUT
White
GND
Green
MALE (P#)
NEUT
White
Black
HOT
120 V
GND
Green
As viewed from wiring side

CIRCUIT INTERRUPTERS
CIRCUIT BREAKER (CB#)
FUSE (F#)

MISC
METER (M#)
ANTENNA
* = V, mV, A, mA, mA
LAMPS (DS#)
Incandescent
NEON

arrl_symbols04

Figure T6-1 — Questions in the following sections ask you to identify specific schematic symbols, and you can use this figure to identify each of those symbols. Circuits T1, T2 and T3 are constructed entirely from symbols defined in this figure.

8   Subelement T6

Figure T1

T6C02   What is component 1 in figure T1?

A. Resistor
B. Transistor
C. Battery
D. Connector

A   See Figure T6-1. [*Ham Radio License Manual*, page 3-12]

T6C03   What is component 2 in figure T1?

A. Resistor
B. Transistor
C. Indicator lamp
D. Connector

B   See Figure T6-1. [*Ham Radio License Manual*, page 3-12]

T6C04   What is component 3 in figure T1?

A. Resistor
B. Transistor
C. Lamp
D. Ground symbol

C   See Figure T6-1. [*Ham Radio License Manual*, page 3-12]

T6C05   What is component 4 in figure T1?

A. Resistor
B. Transistor
C. Battery
D. Ground symbol

C   See Figure T6-1. [*Ham Radio License Manual*, page 3-12]

ARRL0558   Figure T2

T6C06   What is component 6 in figure T2?

  A. Resistor
  B. Capacitor
  C. Regulator IC
  D. Transistor

B   See Figure T6-1. [*Ham Radio License Manual*, page 3-12]

T6C07   What is component 8 in figure T2?

  A. Resistor
  B. Inductor
  C. Regulator IC
  D. Light emitting diode

D   See Figure T6-1. [*Ham Radio License Manual*, page 3-12]

T6C08   What is component 9 in figure T2?

  A. Variable capacitor
  B. Variable inductor
  C. Variable resistor
  D. Variable transformer

C   See Figure T6-1. [*Ham Radio License Manual*, page 3-12]

T6C09   What is component 4 in figure T2?

  A. Variable inductor
  B. Double-pole switch
  C. Potentiometer
  D. Transformer

D   See Figure T6-1. [*Ham Radio License Manual*, page 3-12]

Figure T3

T6C10   **What is component 3 in figure T3?**

A. Connector
B. Meter
C. Variable capacitor
D. Variable inductor

D   See Figure T6-1. [*Ham Radio License Manual*, page 3-12]

T6C11   **What is component 4 in figure T3?**

A. Antenna
B. Transmitter
C. Dummy load
D. Ground

A   See Figure T6-1. [*Ham Radio License Manual*, page 3-12]

T6C12   **What do the symbols on an electrical circuit schematic diagram represent?**

A. Electrical components
B. Logic states
C. Digital codes
D. Traffic nodes

A   See question T6C01. [*Ham Radio License Manual*, page 3-12]

ARRL0527

Input ON

DS1

Q1

R1

J1
Input

BT1

Exam Diagram T1

(A)

POWER SUPPLY

F1   S1   D1   DC Out

R2

T1

P1
120 V ac

C1   R1

DS1   D2

Exam Diagram T2

(B)

Ant

ANTENNA TUNER

C1   C2

Input

L1

Exam Diagram T3

(C)

Figure T6-2 — Here are Figures T1 (A), T2 (B) and T3 (C) used on the Technician exam with component designators drawn in. Different letters represent each type of component, for example R for resistor (see Figure T6-1). Lines and dots show electrical connections between the components, but may not correspond to actual wires.

T6C13 **Which of the following is accurately represented in electrical circuit schematic diagrams?**

A. Wire lengths
B. Physical appearance of components
C. The way components are interconnected
D. All of these choices are correct

C A schematic does not illustrate the actual physical layout of a circuit. (A *pictorial diagram* is used for that purpose.) It only shows how the components are connected electrically. The lines drawn from component to component represent those electrical connections. Each line does not necessarily correspond to a physical wire — it just indicates that an electrical connection exists between whatever is at each end of the line. [*Ham Radio License Manual*, page 3-12]

T6D   Component functions

T6D01 **Which of the following devices or circuits changes an alternating current into a varying direct current signal?**

A. Transformer
B. Rectifier
C. Amplifier
D. Reflector

B Heavy-duty diodes that can withstand large voltages and currents are called *rectifiers*. If an ac voltage is applied to a diode, the result is a unidirectional, pulsing dc current because current is blocked when the voltage tries to push electrons in the "wrong" direction. (See question T6B02.) Circuits that change ac current into dc are also called rectifiers. [*Ham Radio License Manual*, page 3-10]

T6D02 **What best describes a relay?**

A. A switch controlled by an electromagnet
B. A current controlled amplifier
C. An optical sensor
D. A pass transistor

A A switch is operated manually while a *relay* is a switch controlled by an electromagnet. [*Ham Radio License Manual*, page 3-12]

T6D03 What type of switch is represented by item 3 in figure T2?

A. Single-pole single-throw
B. Single-pole double-throw
C. Double-pole single-throw
D. Double-pole double-throw

A Figure T2 is shown with question T6C06. Use the symbol chart shown in Figure T6-1 as for questions T6C02 through T6C11. [*Ham Radio License Manual*, page 3-12]

T6D04 Which of the following can be used to display signal strength on a numeric scale?

A. Potentiometer
B. Transistor
C. Meter
D. Relay

C *Indicators* and *displays* are important components for radio equipment. An indicator is either ON or OFF, such as a power indicator or a label that appears when you are transmitting. A *meter* provides information as a value in the form of numbers or on a numeric scale. [*Ham Radio License Manual*, page 3-12]

T6D05 What type of circuit controls the amount of voltage from a power supply?

A. Regulator
B. Oscillator
C. Filter
D. Phase inverter

A A power supply's output voltage changes with the amount of output current. The percentage of voltage change between zero current (no load) and maximum current (full load) is the *regulation* of the supply. To achieve "tight" regulation, meaning little variation as current changes, requires a *regulator* circuit in the supply. [*Ham Radio License Manual*, page 5-14]

T6D06 What component is commonly used to change 120V AC house current to a lower AC voltage for other uses?

A. Variable capacitor
B. Transformer
C. Transistor
D. Diode

B *Transformers* are made from two or more inductors that share their stored energy. This allows energy to be transferred from one inductor to another while changing the combination of voltage and current. A transformer is used to transfer energy from a home's 120 V ac outlets to a lower voltage for use in electronic equipment. [*Ham Radio License Manual*, page 3-8]

T6D07 Which of the following is commonly used as a visual indicator?

A. LED
B. FET
C. Zener diode
D. Bipolar transistor

A See question T6D04. [*Ham Radio License Manual*, page 3-10]

T6D08 Which of the following is used together with an inductor to make a tuned circuit?

A. Resistor
B. Zener diode
C. Potentiometer
D. Capacitor

D Capacitors and inductors are used to create *tuned circuits*. A tuned circuit acts as a filter, passing or rejecting signals at its *resonant* frequency. Tuned circuits are important in radio because they help radios generate or receive signals at a single frequency. [*Ham Radio License Manual*, page 3-9]

T6D09 What is the name of a device that combines several semiconductors and other components into one package?

A. Transducer
B. Multi-pole relay
C. Integrated circuit
D. Transformer

C An *integrated circuit* (*IC* or *chip*) is made of many components connected together as a useful circuit and packaged as a single component. ICs range from very simple circuits consisting of a few diodes all the way to complex microprocessors or signal-processing chips with many thousands of components. [*Ham Radio License Manual*, page 3-10]

T6D10   What is the function of component 2 in Figure T1?

    A. Give off light when current flows through it
    B. Supply electrical energy
    C. Control the flow of current
    D. Convert electrical energy into radio waves

C   Figure T1 is shown with question T6C06. Component 2 is a transistor that can act as a switch or amplifier. (See question T6B01 and Figure T6-1.) [*Ham Radio License Manual*, page 3-10]

T6D11   Which of the following is a common use of coaxial cable?

    A. Carry dc power from a vehicle battery to a mobile radio
    B. Carry RF signals between a radio and antenna
    C. Secure masts, tubing, and other cylindrical objects on towers
    D. Connect data signals from a TNC to a computer

B   Coaxial cable is specially made to carry RF signals without allowing them to radiate or be affected by other signals. Materials that make up coaxial cable are chosen to minimize the amount of signal power lost as the signal travels through the cable. This makes coaxial cable an ideal choice to carry RF signals between a radio and an antenna. [*Ham Radio License Manual*, page 4-8]

Station Equipment

SUBELEMENT T7 — Station equipment; common transmitter and receiver problems, antenna measurements and troubleshooting, basic repair and testing [4 Exam Questions — 4 Groups]

T7A Station radios; receivers, transmitters, transceivers

T7A01 What is the function of a product detector?

A. Detect phase modulated signals
B. Demodulate FM signals
C. Detect CW and SSB signals
D. Combine speech and RF signals

C Detectors for CW and SSB need to reproduce the signal's frequencies, not just their amplitude. To recover these signals, a *product detector* detects CW and SSB signals by combining a signal at the frequency of the modulated signal's RF carrier with the modulated signal itself. The product detector works very much like a mixer except that one of its output products is the original modulating signal, thus the name product detector. [*Ham Radio License Manual*, page 3-17]

Exam Diagram T6   ARRL0529

Figure T7-1 — The superheterodyne receiver uses mixers to shift the receiving signal one or more times. After each shift or conversion, an intermediate frequency (IF) amplifier increases the signal's strength and a filter removes unwanted signals. This is diagram T6 on the Technician exam with all blocks labeled.

T7A02   **What type of receiver is shown in Figure T6?**

    A. Direct conversion
    B. Super-regenerative
    C. Single-conversion superheterodyne
    D. Dual-conversion superheterodyne

C   The *superheterodyne receiver* shown in the figure is the standard in Amateur Radio today. The key characteristic of the "*superhet*" is the use of *mixers* to shift incoming signals of any frequency to a single fixed frequency, called the *intermediate frequency* (*IF*). Unwanted signals can then rejected by narrow filters. An IF amplifier then amplifies the signal so that it can be demodulated and the information recovered. The IF may be either the sum or difference of the frequencies of the oscillator and the modulated RF signal. After amplification and filtering at the IF, the signal is then demodulated. In the figure, a product detector follows the IF amplifier, recovering a CW or SSB signal (see question T7A01). In a superhet, the oscillator used by the product detector is called a *beat frequency oscillator* (BFO) because it can be adjusted to produce different tones or frequencies, called "beats" originally, in the output signal. Because this receiver has only one mixer and IF it is a *single-conversion superheterodyne*. [*Ham Radio License Manual*, page 3-18]

T7A03   **What is the function of a mixer in a superheterodyne receiver?**

    A. To reject signals outside of the desired passband
    B. To combine signals from several stations together
    C. To shift the incoming signal to an intermediate frequency
    D. To connect the receiver with an auxiliary device, such as a TNC

C   See question T7A02. [*Ham Radio License Manual*, page 3-18]

T7A04 **What circuit is pictured in Figure T7, if block 1 is a frequency discriminator?**

A. A double-conversion receiver
B. A regenerative receiver
C. A superheterodyne receiver
D. An FM receiver

D Figure T7-2 (exam diagram T7) shows how a superheterodyne receiver is built to receive an FM voice signal. The input signal passes through the mixer-oscillator-IF amplifier sequence across the top of the diagram. Instead of immediately demodulating the FM signal, however, it is applied to a *limiter*. A limiter is a high-gain amplifier that removes all amplitude variations from the signal, leaving only the frequency variations that represent the modulation of the signal. The output of the limiter can then be applied to a *discriminator* to recover the voice signal that is then amplified and heard on a speaker. [*Ham Radio License Manual*, page 3-19]

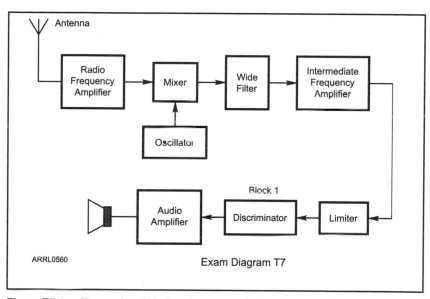

Figure T7-2 — To receive FM signals, a superheterodyne receiver follows the IF stage with a limiter amplifier that strips away all of the amplitude variations caused by noise. This leaves only the frequency variations which are then turned back into speech or data by a frequency discriminator. This is diagram T7 on the Technician exam with all blocks labeled.

Figure T7-3 — A simple CW transmitter consists of an oscillator to generate a low-power signal and two amplifiers to increase the signal strength to a useful level. A telegraph key is used to turn the amplifiers on and off, creating the CW signal that is transmitted. The oscillator runs continuously so that its frequency remains stable. This is diagram T4 on the Technician exam with all blocks labeled.

T7A05 What is the function of block 1 if figure T4 is a simple CW transmitter?

A. Reactance modulator
B. Product detector
C. Low-pass filter
D. Oscillator

D Figure T7-3 (exam diagram T4) shows the block diagram of a basic CW transmitter. The *oscillator* at left produces a low-power signal at one frequency. Oscillators are used in both receivers and transmitters to determine the operating frequency. In a transmitter, the output signal from an oscillator is modulated and amplified before actually being applied to an antenna.

The output signal from the oscillator is not strong enough for reliable communication over long distances. The next stage of the transmitter, reading left to right, is an amplifier circuit called a *driver* that makes the signal stronger. The driver stage allows the oscillator to operate continuously at low power so that its frequency remains stable. The output of the driver is then applied to a *power amplifier* whose output is strong enough to provide a useful signal for communication with other stations.

In order to turn the output signal on and off as Morse code, the driver and power amplifier stages are keyed (switched on and off) by a telegraph key. The oscillator can then run continuously and that also helps keep its signal frequency stable. [*Ham Radio License Manual*, page 3-15]

T7A06 **What device takes the output of a low-powered 28 MHz SSB exciter and produces a 222 MHz output signal?**

A. High-pass filter
B. Low-pass filter
C. Transverter
D. Phase converter

C The superheterodyne technique of shifting signal frequencies can also be applied at the equipment level to convert an entire transceiver to operate on a band other than it was designed for. Instead of a mixer, a piece of equipment called a *transverter* is used. Low-power transmitter output signals on one band are shifted to the new output frequency where they are amplified for transmission. A receiving converter (basically a mixer) shifts incoming signals to the desired band where they are received as regular signals by the transceiver. Transverters are used by many hams to allow one main transceiver to be used on one or more new bands. For example, with few transceivers available for CW and SSB operation on 222 MHz, a transverter is used to convert 222 MHz signals to and from the 28 MHz band available on all HF gear. [*Ham Radio License Manual*, page 3-19]

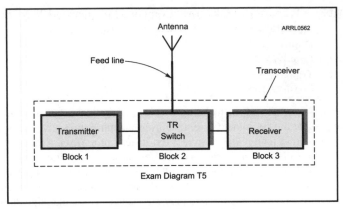

Figure T7-4 — A basic radio station is made up of a transmitter and receiver connected to an antenna with a feed line. The transmit-receive (TR) switch allows the transmitter and receiver to share the antenna. A transceiver includes all of those pieces in a single enclosure. This is diagram T5 on the Technician exam with all blocks labeled.

T7A07   **If figure T5 represents a transceiver in which block 1 is the transmitter portion and block 3 is the receiver portion, what is the function of block 2?**

A. A balanced modulator
B. A transmit-receive switch
C. A power amplifier
D. A high-pass filter

B   The three basic elements of a ham radio station, big or small, are the transmitter, receiver, and antenna as shown in Figure T7-4 (exam diagram T5). A transmitter (abbreviated XMTR) generates a signal that carries speech, Morse code or data information. A receiver (abbreviated RCVR) recovers the speech, Morse code or data information from a signal. An antenna turns the radio signals from a transmitter into energy that travels through space as a radio wave. An antenna also captures radio waves and turns them into signals for a receiver to work with. A feed line connects the antenna to the transmitter or receiver. Feed lines are also called transmission lines, just like power lines, because they are used to transfer energy — radio signals in this case. Most amateur equipment combines the transmitter and receiver into a single piece of equipment called a transceiver (abbreviated XCVR). This is what hams mean when they refer to something as a radio. A transceiver shares a single antenna between the transmitter and receiver circuits by using a *transmit-receive (TR) switch*. [*Ham Radio License Manual*, page 2-11]

T7A08   Which of the following circuits combines a speech signal and an RF carrier?

A. Beat frequency oscillator
B. Discriminator
C. Modulator
D. Noise blanker

C   The process of combining data or voice signals with an RF signal is *modulation*. A circuit that performs the modulation function is therefore called a *modulator*. The function of the modulator is to add the data or voice signal to an RF signal or *carrier*. The result is an RF signal that can be communicated by radio. A modulator can be as simple as an on-off switch (a telegraph key, for example) or it can be very complex in the case of a high-speed data transmission. [*Ham Radio License Manual*, page 3-16]

T7A09   Which of the following devices is most useful for VHF weak-signal communication?

A. A quarter-wave vertical antenna
B. A multi-mode VHF transceiver
C. An omni-directional antenna
D. A mobile VHF FM transceiver

B   Most VHF "weak-signal" contacts are made using SSB or CW because those modes give better performance over long distances. This style of operating also uses horizontally-polarized directional antennas (beams). Since FM is not used, you'll need a multi-mode transceiver to use SSB or CW for VHF/UHF DXing. [*Ham Radio License Manual*, page 6-30]

T7A10   What device increases the low-power output from a handheld transceiver?

A. A voltage divider
B. A RF power amplifier
C. An impedance network
D. A voltage regulator

B   Amplifiers are used to increase the strength of audio or radio signals. Amplifiers used to increase the power of transmitted RF signals are called *power amplifiers* or *linear amplifiers*. Amplifiers used to increase the strength of a received signal are called *preamplifiers* or just *preamps*. [*Ham Radio License Manual*, page 5-7]

T7A11 Which of the following circuits demodulates FM signals?

A. Limiter
B. Discriminator
C. Product detector
D. Phase inverter

B   See question T7A04. [*Ham Radio License Manual*, page 3-17]

T7A12 Which term describes the ability of a receiver to discriminate between multiple signals?

A. Tuning rate
B. Sensitivity
C. Selectivity
D. Noise floor

C   Receivers are compared on the basis of two primary characteristics; *sensitivity* and *selectivity*. A receiver's sensitivity determines its ability to detect weak signals. Higher sensitivity means a receiver can detect weaker signals. Sensitivity is specified as a *minimum detectable signal* level, usually in microvolts. Selectivity is the ability of a receiver to retrieve the information from just the desired signal in the presence of unwanted signals. High selectivity means that a receiver can operate properly even in the presence of strong signals on nearby frequencies. Receivers use filters to reject the unwanted signals. [*Ham Radio License Manual*, page 3-19]

T7A13 Where is an RF preamplifier installed?

A. Between the antenna and receiver
B. At the output of the transmitter's power amplifier
C. Between a transmitter and antenna tuner
D. At the receiver's audio output

A   If a receiver is not sensitive enough (see question T7A12), a *preamplifier* or *preamp* can be connected between the antenna and receiver. [*Ham Radio License Manual*, page 3-19]

T7B Common transmitter and receiver problems; symptoms of overload and overdrive, distortion, interference, over and under modulation, RF feedback, off frequency signals; fading and noise; problems with digital communications interfaces

T7B01 What can you do if you are told your FM handheld or mobile transceiver is over deviating?

A. Talk louder into the microphone
B. Let the transceiver cool off
C. Change to a higher power level
D. Talk farther away from the microphone

D Excessive modulation for all types of speech results in distortion of transmitted speech and unwanted or *spurious* transmitter outputs on adjacent frequencies where they cause interference. Those unwanted transmitter outputs have lots of names, but the most common is *splatter*. Generating those outputs is called *splattering*, as in, "You're splattering 10 kHz away!" An overmodulated FM signal has excessive deviation and is said to be *over-deviating*. Over-deviation is usually caused by speaking too loudly into the microphone and may cause interference on adjacent channels. An FM transmitter can also be internally misadjusted to over-deviate at normal speech levels. To reduce over-deviation, speak more softly or move the microphone farther from your mouth. [*Ham Radio License Manual*, page 5-4]

T7B02 What is meant by fundamental overload in reference to a receiver?

A. Too much voltage from the power supply
B. Too much current from the power supply
C. Interference caused by very strong signals
D. Interference caused by turning the volume up too high

C Very strong signals may overwhelm a receiver's ability to reject them. Receiver overload is a common type of interference to TV and FM-broadcast receivers. It happens most often to consumer electronic equipment near an amateur station or other transmitter. Symptoms of overload include severe interference on all channels of a TV or FM receiver or an amateur may hear noise across an entire band when the strong signal is present. If adding attenuation (either by turning on a receiver's attenuator or removing an antenna) causes the interference to disappear, it's probably caused by overload. When the RF signal (at the fundamental frequency) enters the receiver, it overloads one or more circuits. The receiver front end (first circuit stages after the antenna) is most commonly affected. [*Ham Radio License Manual*, page 5-20]

T7B03   Which of the following may be a cause of radio frequency interference?

A. Fundamental overload
B. Harmonics
C. Spurious emissions
D. All of these choices are correct

D   All three of the choices cause interference by disrupting receiver performance (A) or by creating unwanted signals that directly interfere with desired signals (B and C). As more and more electronic devices and electrical appliances are put in use every day, interference between them and radios, called radio frequency interference (RFI), becomes more commonplace. RFI can occur in either "direction" — to or from the Amateur Radio equipment. Interference becomes more severe with higher power or closer spacing to the signal source. The ARRL's Technical Information Service Web page provides information on all kinds of RFI and the means to correct it. [*Ham Radio License Manual*, page 5-18]

T7B04   What is the most likely cause of interference to a non-cordless telephone from a nearby transmitter?

A. Harmonics from the transmitter
B. The telephone is inadvertently acting as a radio receiver
C. Poor station grounding
D. Improper transmitter adjustment

B   The major cause of telephone interference comes from telephones that were not equipped with interference protection when they were manufactured. (Reference: FCC CIB Telephone Interference Bulletin) Radio frequency energy from your amateur transmitter may be strong enough to cause the telephone to act as a receiver through *direct detection*, the most common form of interference to telephones. As with receiver overload, there is nothing you can do at the transmitter to cure the interference. Interference protection measures must be taken at, or in, the telephone in question. [*Ham Radio License Manual*, page 5-20]

T7B05 What is a logical first step when attempting to cure a radio frequency interference problem in a nearby telephone?

A. Install a low-pass filter at the transmitter
B. Install a high-pass filter at the transmitter
C. Install an RF filter at the telephone
D. Improve station grounding

C If a telephone is responding to a strong nearby RF signal, the first step should be to prevent the RF from getting into the telephone in the first place. Since RF is probably entering the telephone through the wiring that connects it to the telephone network, an RF filter should be used. Low-pass filters connected at the telephone's modular jack are the best way to reduce RFI from direct detection. These filters are available at electronics stores. Cordless phones may also suffer from fundamental overload as described in question T7B04. [*Ham Radio License Manual*, page 5-20]

T7B06 What should you do first if someone tells you that your station's transmissions are interfering with their radio or TV reception?

A. Make sure that your station is functioning properly and that it does not cause interference to your own television
B. Immediately turn off your transmitter and contact the nearest FCC office for assistance
C. Tell them that your license gives you the right to transmit and nothing can be done to reduce the interference
D. Continue operating normally because your equipment cannot possibly cause any interference

A You should first make sure that your equipment is operating properly and not creating spurious signals on the TV channels. Check for interference to your own TV. If you see it, stop operating and cure the problem before you go back on the air. [*Ham Radio License Manual*, page 5-21]

T7B07 Which of the following may be useful in correcting a radio frequency interference problem?

A. Snap-on ferrite chokes
B. Low-pass and high-pass filters
C. Band-reject and band-pass filters
D. All of these choices are correct

D Knowing how to use the different kinds of filters and RF blocking materials such as ferrite cores is an important part of the amateur's repertoire. If you are having an interference problem, members of your club may be able to help you. The ARRL also maintains an RF Interference resource Web page as part of the Technical Information Service area on the ARRL Web site. [*Ham Radio License Manual*, page 5-18]

T7B08　What should you do if a "Part 15" device in your neighbor's home is causing harmful interference to your amateur station?

A. Work with your neighbor to identify the offending device
B. Politely inform your neighbor about the rules that require him to stop using the device if it causes interference
C. Check your station and make sure it meets the standards of good amateur practice
D. All of these choices are correct

D　Part 15 refers to the section of the FCC Rules that permits unlicensed devices to use radio frequencies as part of their function. This includes cordless phones, baby monitors, audio and video relay devices, and wireless computer data links. These devices are allowed to use the radio spectrum, but are not permitted to cause interference to stations in licensed services, such as Amateur Radio. This is one of the benefits of being a licensed station — protection against interference. Users of Part 15 devices are required to accept interference to the devices from stations in a licensed service, such as Amateur Radio or a broadcast station. These rules must be printed either directly on the device or in the owner's manual supplied with the device. While this means that the responsibility for stopping the interference may lie with your neighbor, you should still make sure the interference is not occurring because of some problem in your station. If you are sure your station is operating properly, you may need to politely educate your neighbor about the interference and help identify the offending device. [*Ham Radio License Manual*, page 5-22]

T7B09　What could be happening if another operator reports a variable high-pitched whine on the audio from your mobile transmitter?

A. Your microphone is picking up noise from an open window
B. You have the volume on your receiver set too high
C. You need to adjust your squelch control
D. Noise on the vehicle's electrical system is being transmitted along with your speech audio

D　Vehicle power wiring often carries a significant amount of noise that can affect your radio's operation. Alternator whine is caused by noise on the dc power system inside your own vehicle. You might hear it with the received audio but more likely it will be heard by others as a high-pitched whine on your transmitted audio that varies with your engine speed. It can be removed by a dc power filter at your radio. [*Ham Radio License Manual*, page 5-15]

T7B10 **What might be the problem if you receive a report that your audio signal through the repeater is distorted or unintelligible?**

A. Your transmitter may be slightly off frequency
B. Your batteries may be running low
C. You could be in a bad location
D. All of these choices are correct

D A station that is slightly off-frequency will be strong, but distorted. This sometimes happens when a radio control key gets bumped, changing frequency by a small amount. If accidentally pressing a control key on your radio is a frequent problem (smaller radios are particularly prone to it), try using the LOCK feature of your radio to disable unintended key presses. You could also be causing excessive deviation by speaking too loudly into the microphone. Either lower your voice or hold the microphone farther away from your mouth. Weak or low batteries can also cause distorted audio by causing the transmitter's modulator circuitry to function improperly. [*Ham Radio License Manual*, page 6-12]

T7B11 **What is a symptom of RF feedback in a transmitter or transceiver?**

A. Excessive SWR at the antenna connection
B. The transmitter will not stay on the desired frequency
C. Reports of garbled, distorted, or unintelligible transmissions
D. Frequent blowing of power supply fuses

C It is not unusual for RF current flowing in sensitive audio cables or data cables to interfere with your station's normal function, just as your strong transmitter signal might be picked up and detected by a neighbor's telephone or audio system. "RF feedback" via a microphone cable can cause distorted transmitted audio, for example. [*Ham Radio License Manual*, page 5-23]

T7B12 **What does the acronym "BER" mean when applied to digital communications systems?**

A. Baud Enhancement Recovery
B. Baud Error Removal
C. Bit Error Rate
D. Bit Exponent Resource

C A radio signal experiences many disruptions between the transmitter and receiver; fading, interference, noise often introduce errors into the stream of data, measured by the *bit error rate* (*BER*). These would cause errors in the received information if no precautions were taken. Some codes include extra data elements to allow the receiver to detect an error. For example, a *parity bit* can be used to detect simple errors in a single character of data. Some protocols add special data so that common errors can be detected and sometimes even corrected. [*Ham Radio License Manual*, page 5-8]

T7C   Antenna measurements and troubleshooting; measuring SWR, dummy loads, feedline failure modes

T7C01   What is the primary purpose of a dummy load?

A. To prevent the radiation of signals when making tests
B. To prevent over-modulation of your transmitter
C. To improve the radiation from your antenna
D. To improve the signal to noise ratio of your receiver

A   To avoid interfering with other stations while you're adjusting your transmitter or measuring its output power, it's a good idea to use a *dummy load*. A dummy load is a heavy-duty resistor that can absorb and dissipate the output power from a transmitter. Because the dummy load absorbs all of the transmitter output and radiates it as heat instead of radio waves, there is no signal to interfere with other hams. You should always use a dummy load when testing a transmitter unless you absolutely have to make an on-the-air test. [*Ham Radio License Manual*, page 5-4]

T7C02   Which of the following instruments can be used to determine if an antenna is resonant at the desired operating frequency?

A. A VTVM
B. An antenna analyzer
C. A "Q" meter
D. A frequency counter

B   An *antenna analyzer* consists of a very low-power signal source with an adjustable frequency and one or more meters or displays to show the impedance and SWR. It is used to measure an antenna system without using a transmitter whose signal might cause interference. [*Ham Radio License Manual*, page 4-18]

T7C03   What, in general terms, is standing wave ratio (SWR)?

A. A measure of how well a load is matched to a transmission line
B. The ratio of high to low impedance in a feedline
C. The transmitter efficiency ratio
D. An indication of the quality of your station's ground connection

A   *Standing wave ratio*, or SWR, is caused by a mismatch of the characteristic impedance of the feed line and that of a load (such as an antenna) attached to the feed line. Some of the energy delivered by the feed line is reflected at the load and interferes with the incoming energy, creating stationary patterns called *standing waves*. The magnitude of the SWR is the ratio of the maximum and minimum values of the standing wave pattern. SWR is also the ratio of the higher of the two impedances to the lower. [*Ham Radio License Manual*, page 4-10]

T7C04   What reading on an SWR meter indicates a perfect impedance match between the antenna and the feedline?

A. 2 to 1
B. 1 to 3
C. 1 to 1
D. 10 to 1

C   When there is no reflected power there is no interference pattern and the SWR is 1:1. This condition is called a *perfect match*. [*Ham Radio License Manual*, page 4-10]

T7C05   What is the approximate SWR value above which the protection circuits in most solid-state transmitters begin to reduce transmitter power?

A. 2 to 1
B. 1 to 2
C. 6 to 1
D. 10 to 1

A   One possible effect of high SWR is that the interference pattern causes voltages to increase in the feed line and at the transmitter's output where the feed line is connected. The higher voltages can damage a transmitter's output circuits. Most amateur transmitting equipment is designed to work at full power with an SWR of 2:1 or lower. SWR greater than 2:1 may cause the transmitter to reduce power automatically in order to protect its output circuit. [*Ham Radio License Manual*, page 4-10]

T7C06   What does an SWR reading of 4:1 mean?

A. An antenna loss of 4 dB
B. A good impedance match
C. An antenna gain of 4
D. An impedance mismatch

D   SWR greater than 1:1 is called an *impedance mismatch* or *mismatch*. [*Ham Radio License Manual*, page 4-10]

T7C07   **What happens to power lost in a feedline?**

A. It increases the SWR
B. It comes back into your transmitter and could cause damage
C. It is converted into heat
D. It can cause distortion of your signal

C   As energy travels through a feed line, some of it is absorbed by the insulation in the line and some is lost in the resistance of the conductors themselves. These losses are in the form of heat, just as if the energy was dissipated by a resistor. Feed lines used at radio frequencies use special materials and construction methods to minimize power being dissipated as heat by feed line loss and to avoid signals leaking in or out. [*Ham Radio License Manual*, page 4-8]

T7C08   **What instrument other than an SWR meter could you use to determine if a feedline and antenna are properly matched?**

A. Voltmeter
B. Ohmmeter
C. Iambic pentameter
D. Directional wattmeter

D   A directional wattmeter can tell you how much energy is flowing in each direction along a feed line. From the relative amounts of power flowing in each direction, you can calculate SWR. By minimizing reflected power, you can adjust your antenna for the best impedance match. [*Ham Radio License Manual*, page 4-17]

T7C09   **Which of the following is the most common cause for failure of coaxial cables?**

A. Moisture contamination
B. Gamma rays
C. The velocity factor exceeds 1.0
D. Overloading

A   Coaxial cables must be protected. The performance of coaxial cable depends on the integrity of its outer coating, the jacket. Nicks, cuts and scrapes can all breach the jacket allowing moisture contamination, the most common cause of coaxial cable failure. Prolonged exposure to the ultraviolet (UV) in sunlight will also cause the plastic in the jacket to degrade, causing small cracks that allow water into the cable. To protect the cable against UV damage the jacket usually contains a pigment that absorbs and blocks the UV. Capillary action of the strands making up the braided outer shield can also draw water into coaxial cable. Once in the cable, the water causes both corrosion and heat losses in the shield. Water can also get into coaxial cable through an improperly sealed connection. [*Ham Radio License Manual*, page 4-16]

T7C10 **Why should the outer jacket of coaxial cable be resistant to ultraviolet light?**

A. Ultraviolet resistant jackets prevent harmonic radiation
B. Ultraviolet light can increase losses in the cable's jacket
C. Ultraviolet and RF signals can mix together, causing interference
D. Ultraviolet light can damage the jacket and allow water to enter the cable

D   See question T7C09. [*Ham Radio License Manual*, page 4-16]

T7C11 **What is a disadvantage of "air core" coaxial cable when compared to foam or solid dielectric types?**

A. It has more loss per foot
B. It cannot be used for VHF or UHF antennas
C. It requires special techniques to prevent water absorption
D. It cannot be used at below freezing temperatures

C   Coax connectors exposed to the weather must be carefully waterproofed. If you use low-loss air-core coax, you will need to pay extra attention to waterproofing the connectors because special techniques are required to prevent water absorption by this cable. [*Ham Radio License Manual*, page 4-17]

T7D   Basic repair and testing; soldering, use of a voltmeter, ammeter, and ohmmeter

T7D01 **Which instrument would you use to measure electric potential or electromotive force?**

A. An ammeter
B. A voltmeter
C. A wavemeter
D. An ohmmeter

B   Electromagnetic force (EMF) is another term for *voltage*, which is measured in *volts*. A meter used to measure voltage is called a *voltmeter*. [*Ham Radio License Manual*, page 3-1]

T7D02 **What is the correct way to connect a voltmeter to a circuit?**

A. In series with the circuit
B. In parallel with the circuit
C. In quadrature with the circuit
D. In phase with the circuit

B   Voltmeters are connected in parallel with a component (also termed "across a component") or circuit to measure voltage. Voltage is always measured from one point to another or with respect to some reference voltage. [*Ham Radio License Manual*, page 3-3]

T7D03   How is an ammeter usually connected to a circuit?

    A. In series with the circuit
    B. In parallel with the circuit
    C. In quadrature with the circuit
    D. In phase with the circuit

A   Ammeters are connected in series with a component or circuit to measure current. Some meters calibrated in amperes are really voltmeters that measure the voltage across a small resistor in series with the current. Be sure to determine which type of meter you have before using it. [*Ham Radio License Manual*, page 3-3]

T7D04   Which instrument is used to measure electric current?

    A. An ohmmeter
    B. A wavemeter
    C. A voltmeter
    D. An ammeter

D   The units of current flow are amperes and a meter used to measure current is called an *ammeter*. [*Ham Radio License Manual*, page 3-1]

T7D05   What instrument is used to measure resistance?

    A. An oscilloscope
    B. A spectrum analyzer
    C. A noise bridge
    D. An ohmmeter

D   The units of resistance are ohms and a meter used to measure resistance is called an *ohmmeter*. [*Ham Radio License Manual*, page 3-4]

T7D06   Which of the following might damage a multimeter?

    A. Measuring a voltage too small for the chosen scale
    B. Leaving the meter in the milliamps position overnight
    C. Attempting to measure voltage when using the resistance setting
    D. Not allowing it to warm up properly

C   The flexibility afforded by a multimeter also means that it is important to use the meter properly. Trying to measure voltage or connecting the probes to an energized circuit when the meter is set to measure resistance is a common way to damage a multimeter, for example. [*Ham Radio License Manual*, page 3-3]

T7D07 **Which of the following measurements are commonly made using a multimeter?**

A. SWR and RF power
B. Signal strength and noise
C. Impedance and reactance
D. Voltage and resistance

D The typical multimeter uses a switch and different sets of input connections to select which parameter and range of values to measure. The most common measurements are voltage, resistance, and current. The convenience of having all the different functions in a single instrument more than outweighs the extra complexity of learning to use a multimeter. [*Ham Radio License Manual*, page 3-3]

T7D08 **Which of the following types of solder is best for radio and electronic use?**

A. Acid-core solder
B. Silver solder
C. Rosin-core solder
D. Aluminum solder

C Learning how to install your own coax connectors and make cables not only saves money but allows you to make repairs at home and under emergency conditions! Start by reading the ARRL Technical Information Service's online article "The Art of Soldering." Follow up with "Connectors For All Occasions, Parts 1 and 2" on the same Web page. You'll learn what kind of solder to use for electronics (rosin-core), what a "cold" solder joint looks like (it has a grainy or dull surface), and many other useful tips. [*Ham Radio License Manual*, page 4-17]

T7D09 **What is the characteristic appearance of a "cold" solder joint?**

A. Dark black spots
B. A bright or shiny surface
C. A grainy or dull surface
D. A greenish tint

C See question T7D08. [*Ham Radio License Manual*, page 4-17]

T7D10 **What is probably happening when an ohmmeter, connected across a circuit, initially indicates a low resistance and then shows increasing resistance with time?**

A. The ohmmeter is defective
B. The circuit contains a large capacitor
C. The circuit contains a large inductor
D. The circuit is a relaxation oscillator

B This is an indication that a large (high-value) capacitor is part of the circuit. As the meter transfers energy to the capacitor it begins to appear more like a high-value resistor to the multimeter's dc resistance measuring circuit. You'll find out more about capacitor charging when you begin studying for your General class license. [*Ham Radio License Manual*, page 3-3]

T7D11 **Which of the following precautions should be taken when measuring circuit resistance with an ohmmeter?**

A. Ensure that the applied voltages are correct
B. Ensure that the circuit is not powered
C. Ensure that the circuit is grounded
D. Ensure that the circuit is operating at the correct frequency

B A circuit with power applied can cause enough current to flow in a multimeter's sensitive resistance measurement circuits to cause measurement errors or even damage the meter. Always make sure power is removed from a circuit before making resistance measurements. [*Ham Radio License Manual*, page 3-3]

T8

Operating Modes and Special Operations

SUBELEMENT T8 — Modulation modes; amateur satellite operation, operating activities, non-voice communications [4 Exam Questions — 4 Groups]

T8A   Modulation modes; bandwidth of various signals

T8A01   Which of the following is a form of amplitude modulation?

A. Spread-spectrum
B. Packet radio
C. Single sideband
D. Phase shift keying

C   An *amplitude modulation* (*AM*) signal consists of a carrier and two sidebands; one higher in frequency than the carrier (the upper sideband, USB) and one lower (the lower sideband, LSB). A *single-sideband* (SSB) signal is created by removing the carrier and one of the sidebands. [*Ham Radio License Manual*, page 2-8]

T8A02   What type of modulation is most commonly used for VHF packet radio transmissions?

A. FM
B. SSB
C. AM
D. Spread Spectrum

A   FM can be used for data signals, such as those for packet radio on VHF and UHF. The data is sent as audio tones by using the FM transmitter voice input and speaker output. This allows inexpensive FM voice radios to be used. [*Ham Radio License Manual*, page 2-10]

T8A03 Which type of voice modulation is most often used for long-distance or weak signal contacts on the VHF and UHF bands?

A. FM
B. AM
C. SSB
D. PM

C Because the SSB signal's power is concentrated into a narrow bandwidth, it is possible to communicate with SSB over much longer ranges and in poorer conditions than with FM or AM, particularly on the VHF and UHF bands. That is why the VHF and UHF DXers and contest operators use SSB signals. [*Ham Radio License Manual*, page 2-10]

T8A04 Which type of modulation is most commonly used for VHF and UHF voice repeaters?

A. AM
B. SSB
C. PSK
D. FM

D *Frequency modulation (FM)* and its close relative *phase modulation (PM)* are used in repeater communication because of the mode's superior rejection of noise and static. [*Ham Radio License Manual*, page 2-10]

T8A05 Which of the following types of emission has the narrowest bandwidth?

A. FM voice
B. SSB voice
C. CW
D. Slow-scan TV

C CW signals consist of a continuous wave signal on a single frequency that is turned on and off in the patterns that make up the Morse code. As a result, CW has the narrowest bandwidth of any mode copied directly by the human ear. [*Ham Radio License Manual*, page 2-10]

T8A06 Which sideband is normally used for 10 meter HF, VHF and UHF single-sideband communications?

A. Upper sideband
B. Lower sideband
C. Suppressed sideband
D. Inverted sideband

A There is no technical reason for choosing USB over LSB. However, in order to make communications easier, amateur radio has standardized on the following conventions:

• Below 10 MHz, LSB is used

• Above 10 MHz, USB is used — including all of the VHF and UHF bands

This convention is even programmed into radio equipment as the normal operating mode! There is one exception: amateurs are required to use USB on the five 60 meter band (5 MHz) channels. [*Ham Radio License Manual*, page 2-10]

T8A07 **What is the primary advantage of single sideband over FM for voice transmissions?**

    A. SSB signals arc easier to tune
    B. SSB signals are less susceptible to interference
    C. SSB signals have narrower bandwidth
    D. All of these choices are correct

C Whereas an FM signal may have a bandwidth of 5 to 15 kHz, SSB signals only require 2 to 3 kHz of spectrum. This increases the usable range of SSB signals compared to FM signals of equal power. (See question T8A03.) [*Ham Radio License Manual*, page 2-10]

T8A08 **What is the approximate bandwidth of a single sideband voice signal?**

    A. 1 kHz
    B. 3 kHz
    C. 6 kHz
    D. 15 kHz

B See the following table of Signal Bandwidths for various modes. [*Ham Radio License Manual*, page 2-10]

Signal Bandwidths

Type of Signal	Typical Bandwidth
AM amateur voice	6 kHz
AM broadcast	10 kHz
Commercial video broadcast	6 MHz (see note)
SSB voice	2 to 3 kHz
SSB digital	500 to 3000 Hz (0.5 to 3 kHz)
CW	100 to 300 Hz (0.1 to 0.3 kHz)
FM amateur voice	5 to 15 kHz
FM broadcast	150 kHz

Note: On June 12, 2009 US broadcasters converted all over-the-air TV signals to digital modulation. Within the 6 MHz channel, there may be from four to five digitally-compressed audio-video programs, each with a 1.2-1.5 MHz bandwidth. Amateurs will continue to use the older analog format for fast-scan television for the foreseeable future and the bandwidth of those signals is approximately 6 MHz.

T8A09 What is the approximate bandwidth of a VHF repeater FM phone signal?

A. Less than 500 Hz
B. About 150 kHz
C. Between 5 and 15 kHz
D. Between 50 and 125 kHz

C See the table of Signal Bandwidths. [*Ham Radio License Manual*, page 2-10]

T8A10 What is the typical bandwidth of analog fast-scan TV transmissions on the 70 cm band?

A. More than 10 MHz
B. About 6 MHz
C. About 3 MHz
D. About 1 MHz

B See the table of Signal Bandwidths. [*Ham Radio License Manual*, page 2-10]

T8A11 What is the approximate maximum bandwidth required to transmit a CW signal?

A. 2.4 kHz
B. 150 Hz
C. 1000 Hz
D. 15 kHz

B See the table of Signal Bandwidths. [*Ham Radio License Manual*, page 2-10]

T8B   Amateur satellite operation; Doppler shift, basic orbits, operating protocols

T8B01 Who may be the control operator of a station communicating through an amateur satellite or space station?

A. Only an Amateur Extra Class operator
B. A General Class licensee or higher licensee who has a satellite operator certification
C. Only an Amateur Extra Class operator who is also an AMSAT member
D. Any amateur whose license privileges allow them to transmit on the satellite uplink frequency

D [97.209 (a)] — The control operator of a station communicating through an amateur satellite may hold any class of amateur license that has privileges to transmit in the satellite's input frequency range, regardless of the frequency of the satellite's downlink signal. [*Ham Radio License Manual*, page 6-32]

T8B02 How much transmitter power should be used on the uplink frequency of an amateur satellite or space station?

A. The maximum power of your transmitter
B. The minimum amount of power needed to complete the contact
C. No more than half the rating of your linear amplifier
D. Never more than 1 watt

B Using minimum power is particularly important when using satellites because of their very limited power sources. By transmitting with excessive power, you can actually cause the signals of other stations using the satellite to be reduced as they are relayed by the satellite. [*Ham Radio License Manual*, page 6-33]

T8B03 Which of the following can be done using an amateur radio satellite?

A. Talk to amateur radio operators in other countries
B. Get global positioning information
C. Make telephone calls
D. All of these choices are correct

A Because the satellite is high above the Earth, its transmitted signal can often be heard in many countries at the same time. The satellites farthest from Earth can provide coverage over nearly half of the Earth's surface! [*Ham Radio License Manual*, page 6-32]

T8B04 Which amateur stations may make contact with an amateur station on the International Space Station using 2 meter and 70 cm band amateur radio frequencies?

A. Only members of amateur radio clubs at NASA facilities
B. Any amateur holding a Technician or higher class license
C. Only the astronaut's family members who are hams
D. You cannot talk to the ISS on amateur radio frequencies

B The astronaut-amateurs on board the ISS operate according to the same FCC Rules that earthbound hams do, so there is no reason that communications between them should be restricted based on license class. Any amateur with a license that permits communications on the VHF frequencies used by the ISS is welcome to make contact! [*Ham Radio License Manual*, page 6-32]

T8B05   What is a satellite beacon?

A. The primary transmit antenna on the satellite
B. An indicator light that that shows where to point your antenna
C. A reflective surface on the satellite
D. A transmission from a space station that contains information about a satellite

D   Like its terrestrial counterparts, a satellite's beacon transmits communications so that propagation to the satellite and information about the satellite can be observed. Receiving the beacon from a satellite indicates that it is within range and contacts with it or through it can be attempted. [*Ham Radio License Manual*, page 6-32]

T8B06   What can be used to determine the time period during which an amateur satellite or space station can be accessed?

A. A GPS receiver
B. A field strength meter
C. A telescope
D. A satellite tracking program

D   Satellites are typically only visible for short periods at a time, called a *pass*, because they are moving so rapidly. A computer can perform the calculations required to tell when a particular satellite is visible from your location by using a satellite-tracking program. The program uses mathematical coordinates called *Keplerian elements* available from online Web sites. The output of the program tells you when and where to look in the sky for the satellite. [*Ham Radio License Manual*, page 6-33]

T8B07   With regard to satellite communications, what is Doppler shift?

A. A change in the satellite orbit
B. A mode where the satellite receives signals on one band and transmits on another
C. An observed change in signal frequency caused by the relative motion between the satellite and the earth station
D. A special digital communications mode for some satellites

C   Doppler shift or Doppler effect describes the way the downlink frequency of a satellite varies by several kilohertz during an orbit due to its motion relative to the receiving station. As the satellite is moving toward the receiving station, the frequency of the satellite's signal will increase by a small amount. After the satellite passes overhead, the frequency of the signal will begin to decrease. This is the same effect that causes the sound of a vehicle's horn or whistle to change pitch as it moves past you. [*Ham Radio License Manual*, page 6-32]

T8B08 **What is meant by the statement that a satellite is operating in "mode U/V"?**

A. The satellite uplink is in the 15 meter band and the downlink is in the 10 meter band
B. The satellite uplink is in the 70 cm band and the downlink is in the 2 meter band
C. The satellite operates using ultraviolet frequencies
D. The satellite frequencies are usually variable

B The satellite's operational *mode* specifies the bands on which it is transmitting and receiving. Most satellites only have one mode, but some have several that can be controlled by ground stations. Mode is specified as two letters separated by a slash. The first letter indicates the uplink band and the second letter indicates the downlink band. For example, the uplink for a satellite in U/V mode is in the UHF band (70 cm) and a downlink in the VHF band (2 meters). [*Ham Radio License Manual*, page 6-33]

T8B09 **What causes "spin fading" when referring to satellite signals?**

A. Circular polarized noise interference radiated from the sun
B. Rotation of the satellite and its antennas
C. Doppler shift of the received signal
D. Interfering signals within the satellite uplink band

B Most satellites spin to stabilize their orientation in space. Spin fading is caused by rotation of the satellite and its antennas with respect to the receiving station. As the satellite spins, it's signal will strengthen and weaken depending on the orientation of the antennas. [*Ham Radio License Manual*, page 6-33]

T8B10 **What do the initials LEO tell you about an amateur satellite?**

A. The satellite battery is in Low Energy Operation mode
B. The satellite is performing a Lunar Ejection Orbit maneuver
C. The satellite is in a Low Earth Orbit
D. The satellite uses Light Emitting Optics

C *Low Earth Orbit* is the term for satellites orbiting up to about 1600 miles above the Earth, including the International Space Station. LEO satellites have nearly circular orbits and are only visible from Earth for a few minutes. Higher orbits are referred to as *Medium Earth Orbit*. Farthest out are the *Geosynchronous Earth Orbit* satellites, whose orbital period matches exactly the rotation of the Earth so that the satellite appears to stay above one spot on Earth. [*Ham Radio License Manual*, page 6-33]

T8B11 What is a commonly used method of sending signals to and from a digital satellite?

A. USB AFSK
B. PSK31
C. FM Packet
D. WSJT

C The "Pacsats" are satellites equipped with FM packet radio stations. These satellites can operate as digipeaters and sometimes as bulletin board stations. [*Ham Radio License Manual*, page 6-33]

T8C Operating activities; radio direction finding, radio control, contests, special event stations, basic linking over Internet

T8C01 Which of the following methods is used to locate sources of noise interference or jamming?

A. Echolocation
B. Doppler radar
C. Radio direction finding
D. Phase locking

C There are a number of techniques that allow a radio operator to determine the bearing to a radio transmitter. Using them is called direction finding. Exercises in which hams try to find a hidden transmitter are also called *foxhunts* or *bunny hunts*. In recent years a new type of outdoor radiosport has reached US shores from Europe and Asia — *radio direction finding* or *RDF*. Held as organized events, direction finding is a hybrid of the radio fox hunt using orienteering skills to navigate outdoors with map and compass. The US Amateur Radio Direction Finding organization (**www. ardfusa.com**) is just one of a number of national groups in this worldwide sport, especially popular with teens and young adults. If you are a hiker or camper, then you might be interested in applying your outdoor skills to ARDF. [*Ham Radio License Manual*, page 6-31]

T8C02 Which of these items would be useful for a hidden transmitter hunt?

A. Calibrated SWR meter
B. A directional antenna
C. A calibrated noise bridge
D. All of these choices are correct

B You don't need much in the way of equipment. A portable radio with a signal strength indicator and a handheld or portable directional antenna, such as a small Yagi beam, will work just fine. The point of the exercise is to determine direction to a radio source, the hidden transmitter, and to do so requires a directional antenna. [*Ham Radio License Manual*, page 6-31]

T8C03   **What popular operating activity involves contacting as many stations as possible during a specified period of time?**

A. Contesting
B. Net operations
C. Public service events
D. Simulated emergency exercises

A   A radio contest or *radiosport* event consists of making as many contacts as you can with stations in a targeted area or on specific bands. Contests last for only a few hours or maybe all weekend. They're a great way to improve your station capabilities and operating skills. See the ARRL Web site for an event calendar and links to contest rules. [*Ham Radio License Manual*, page 6-30]

T8C04   **Which of the following is good procedure when contacting another station in a radio contest?**

A. Be sure to sign only the last two letters of your call if there is a pileup calling the station
B. Work the station twice to be sure that you are in his log
C. Send only the minimum information needed for proper identification and the contest exchange
D. All of these choices are correct

C   Check the contest rules and listen to a few contacts during the contest to find out what information is required in the contest exchange. [*Ham Radio License Manual*, page 6-30]

T8C05   **What is a grid locator?**

A. A letter-number designator assigned to a geographic location
B. A letter-number designator assigned to an azimuth and elevation
C. An instrument for neutralizing a final amplifier
D. An instrument for radio direction finding

A   The *Maidenhead Locator System* (named after the city in England where it was developed) divides the Earth's surface into grids organized by latitude and longitude. The designator for each *grid square* consists of two letter and two numbers, such as CN87 or FM13. For example, ARRL Headquarters in Newington, Connecticut is located in grid square FN31. A further two letters can be added for greater precision, such as FN31pq for the precise location of the ARRL station. (Learn more about grid squares at the ARRL Web site.) [*Ham Radio License Manual*, page 6-4]

T8C06   For what purpose is a temporary "1 by 1" format (letter-number-letter) call sign assigned?

A. To designate an experimental station
B. To honor a deceased relative who was a radio amateur
C. For operations in conjunction with an activity of special significance to the amateur community
D. All of these choices are correct

C   Any FCC-licensed amateur or club can also obtain a special 1×1 (1-by-1) call sign such as W3X or K6P for a short-duration special event of significance to the amateur community. These call signs are special because they only have one letter in both the prefix and suffix. Special event call signs are very popular and contacts with them are sought out by hams, worldwide. [*Ham Radio License Manual*, page 7-21]

T8C07   What is the maximum power allowed when transmitting telecommand signals to radio controlled models?

A. 500 milliwatts
B. 1 watt
C. 25 watts
D. 1500 watts

B   Amateurs may transmit telecommand signals with an output power of up to 1 watt. Although the signals do not identify the licensee on the air, radio control (RC) modelers are required to display their name, call sign and address on the RC transmitter. [*Ham Radio License Manual*, page 6-35]

T8C08   What is required in place of on-air station identification when sending signals to a radio control model using amateur frequencies?

A. Voice identification must be transmitted every 10 minutes
B. Morse code ID must be sent once per hour
C. A label indicating the licensee's name, call sign and address must be affixed to the transmitter
D. A flag must be affixed to the transmitter antenna with the station call sign in 1 inch high letters or larger

C   See question T8C07. [*Ham Radio License Manual*, page 6-35]

T8C09   How might you obtain a list of active nodes that use VoIP?

A. From the FCC Rulebook
B. From your local emergency coordinator
C. From a repeater directory
D. From the local repeater frequency coordinator

C   Both the IRLP and EchoLink systems maintain an Internet directory of repeaters participating in their systems. You can browse the directories to find nodes (repeaters and Internet voice server systems) currently on the air. [*Ham Radio License Manual*, page 6-20]

T8C10   How do you select a specific IRLP node when using a portable transceiver?

A. Choose a specific CTCSS tone
B. Choose the correct DSC tone
C. Access the repeater autopatch
D. Use the keypad to transmit the IRLP node ID

D   To initiate an IRLP or Echolink contact, the initiating station must know the repeater control code to request an IRLP connection — this is the ON code. The ON code varies from repeater to repeater and obtaining it may require membership in a club. Once the ON code is entered via your keypad, the four-digit code for the IRLP node — a destination repeater — is entered. The code then directs the repeater controller to establish the Internet link to the desired node. [*Ham Radio License Manual*, page 6-20]

T8C11   What name is given to an amateur radio station that is used to connect other amateur stations to the Internet?

A. A gateway
B. A repeater
C. A digipeater
D. A beacon

A   An Internet gateway is a special kind of digital station that provides a connection to the Internet for data transmitted by Amateur Radio from other station. Most gateways are set up to forward messages. The most common examples are packet radio bulletin board systems (BBS) and the Winlink RMS stations. Messages with a recognized Internet email address can be sent and retrieved over these systems. Another type of gateway provides direct Internet connectivity so that a computer running standard Web browser software can connect to any Internet address. [*Ham Radio License Manual*, page 5-12]

T8D Non-voice communications; image data, digital modes, CW, packet, PSK31

T8D01 Which of the following is an example of a digital communications method?

A. Packet
B. PSK31
C. MFSK
D. All of these choices are correct

D Packet radio uses a version of the computer-to-computer X.25 data exchange protocol that was adapted to suit amateur use becoming the AX.25 protocol. PSK31 is an abbreviation for *phase-shift keying at 31 baud* for low-rate keyboard-to-keyboard communications in a narrow bandwidth. MFSK uses frequency-shift keying of multiple carriers for data communications. Packet radio is generally used on VHF and UHF frequencies while PSK31 and MFSK are primarily used on the HF bands. [*Ham Radio License Manual*, page 5-8]

T8D02 What does the term APRS mean?

A. Automatic Position Reporting System
B. Associated Public Radio Station
C. Auto Planning Radio Set-up
D. Advanced Polar Radio System

A The *Automatic Position Reporting System* (APRS) was created by WB4APR as a way of transmitting GPS location information over Amateur Radio using a system of digipeaters. These relay points forward the position information and call sign to a system of server computers via the Internet. Once the information is stored on the servers Web sites can access the data and show the position of the station on maps in various ways. [*Ham Radio License Manual*, page 5-10]

Figure T8-1 — To use the Automatic Position Reporting System (APRS), the output of a GPS receiver is connected to a packet radio TNC and a VHF radio. Position information and the call sign of the reporting station are then transferred to the ARPS system by an APRS node, either directly or via a digipeater. Station location can then be viewed via the Internet.

T8D03   Which of the following is normally used when sending automatic location reports via amateur radio?

A. A connection to the vehicle speedometer
B. A WWV receiver
C. A connection to a broadcast FM sub-carrier receiver
D. A Global Positioning System receiver

D   See question T8D02. [*Ham Radio License Manual*, page 5-10]

T8D04   What type of transmission is indicated by the term NTSC?

A. A Normal Transmission mode in Static Circuit
B. A special mode for earth satellite uplink
C. An analog fast scan color TV signal
D. A frame compression scheme for TV signals

C NTSC stands for the *National Television System Committee,* which developed the standards that define broadcast fast-scan analog television signals. (For many years, broadcast TV stations in the United States used the NTSC standard for their analog TV broadcasts before they converted to a digital TV signal standard in 2009.) Amateurs adopted the NTSC standard for amateur television (ATV) transmissions. RS-170 is the standard that describes fast-scan video signals before they are sent over the air. These signals can be received on a commercial analog TV receiver equipped with a suitable frequency converter to tune amateur frequencies. The *ARRL Image Communications Handbook* and Technical Information Service pages on the ARRL Web site provide more information on ATV. [*Ham Radio License Manual,* page 6-34]

T8D05   Which of the following emission modes may be used by a Technician Class operator between 219 and 220 MHz?

A. Spread spectrum
B. Data
C. SSB voice
D. Fast-scan television

B 219-220 MHz is a *mode-restricted* sub-band restricted to digital message forwarding only for amateurs of all license classes. Only point-to-point digital message forwarding is allowed between 219 MHz and 220 MHz. Amateurs have access to these frequencies as a secondary allocation, and a number of restrictions apply, as described in Part 97.303(e). Check the FCC Rules [97.305(c)] for limitations on other bands before you begin operations. [*Ham Radio License Manual,* page 7-13]

T8D06   What does the abbreviation PSK mean?

A. Pulse Shift Keying
B. Phase Shift Keying
C. Packet Short Keying
D. Phased Slide Keying

B In phase-shift keying (PSK), the phase of a signal is varied in order to convey information. Amateurs use PSK of an audio signal to transmit information. The audio signal can be transmitted as an AM or FM signal. [*Ham Radio License Manual,* page 5-9]

T8D07   What is PSK31?

A. A high-rate data transmission mode
B. A method of reducing noise interference to FM signals
C. A method of compressing digital television signal
D. A low-rate data transmission mode

D   See question T8D01. [*Ham Radio License Manual*, page 5-9]

T8D08   Which of the following may be included in packet transmissions?

A. A check sum which permits error detection
B. A header which contains the call sign of the station to which the information is being sent
C. Automatic repeat request in case of error
D. All of these choices are correct

D   Packet radio sends data in bursts called *packets*. Each packet consists of a *header* and data. The header contains information about the packet and the call sign of the destination station. The header also includes a *checksum* that allows the receiver to detect errors. If an error is detected, the receiver automatically requests that the packet be retransmitted until the data is received properly. This is called *ARQ* for *automatic repeat request*. [*Ham Radio License Manual*, page 5-9]

T8D09   What code is used when sending CW in the amateur bands?

A. Baudot
B. Hamming
C. International Morse
D. Gray

C   Amateurs use the International Morse code for CW transmissions. [*Ham Radio License Manual*, page 5-8]

T8D10   Which of the following can be used to transmit CW in the amateur bands?

A. Straight Key
B. Electronic Keyer
C. Computer Keyboard
D. All of these choices are correct

D   For sending Morse code, a *key* is used to turn the transmitter output signal on and off. Morse code's dots and dashes are known as the *elements* of the code. When using a straight key, the operator generates the dots and dashes manually. This is called *hand keying*. The J-38 is a common type of straight key and is popular with beginners. Once you are skilled at "the code," you'll want to go faster by using a *keyer*. This electronic device turns contact closures from a Morse paddle into a stream of Morse code elements. A keyer may be a standalone device or it can be built in to a transceiver. A *paddle* is a pair of levers mounted side by side, each having its own set of contacts, one for dots and one for dashes. Keyers and paddles can generate Morse code much faster than by using a straight key. Computer software can also send CW directly from the keyboard by using a keying interface connected to the key input of a radio. [*Ham Radio License Manual*, page 5-5]

T8D11   What is a "parity" bit?

A. A control code required for automatic position reporting
B. A timing bit used to ensure equal sharing of a frequency
C. An extra code element used to detect errors in received data
D. A "triple width" bit used to signal the end of a character

C   Some digital codes include extra data elements to allow the receiver to detect an error. The parity bit is used to tell the receiver whether the number of "1" bits in the character was odd or even. If the received character does not agree with the parity bit, the character can be discarded or a retransmission can be requested. [*Ham Radio License Manual*, page 5-8]

Subelement T9 —

Antennas and Feed Lines

SUBELEMENT T9 — Antennas, feedlines
[2 Exam Questions — 2 Groups]

T9A   Antennas; vertical and horizontal, concept of gain, common portable and mobile antennas, relationships between antenna length and frequency

T9A01   What is a beam antenna?
- A. An antenna built from aluminum I-beams
- B. An omnidirectional antenna invented by Clarence Beam
- C. An antenna that concentrates signals in one direction
- D. An antenna that reverses the phase of received signals

C   The term "beam" is used in the same sense as a flashlight beam. In fact, the reflector of a flashlight acts to concentrate the light bulb's "signal" in one direction, just as a radio frequency beam antenna does. The beam doesn't create more power, only focuses it. Beams can be used to increase signal level at a distant station or to reject interference or noise. [*Ham Radio License Manual*, page 4-14]

T9A02   Which of the following is true regarding vertical antennas?
- A. The magnetic field is perpendicular to the Earth
- B. The electric field is perpendicular to the Earth
- C. The phase is inverted
- D. The phase is reversed

B   The vertical antenna has a pair of reasons for the name. First, the antenna is, in fact, vertically oriented! Second, the vertical orientation of the antenna element means that the antenna is vertically polarized — the electric field component of its radiated waves is vertical or perpendicular to the surface of the Earth. [*Ham Radio License Manual*, page 4-6]

T9A03   Which of the following describes a simple dipole mounted so the conductor is parallel to the Earth's surface?

A. A ground wave antenna
B. A horizontally polarized antenna
C. A rhombic antenna
D. A vertically polarized antenna

B   Like the vertical, a dipole is called a horizontal antenna for two reasons. The antenna element is oriented horizontally and the radiated radio wave is horizontally polarized — its electric field is oriented parallel to the Earth's surface. [*Ham Radio License Manual*, page 4-11]

T9A04   What is a disadvantage of the "rubber duck" antenna supplied with most handheld radio transceivers?

A. It does not transmit or receive as effectively as a full-sized antenna
B. It transmits a circularly polarized signal
C. If the rubber end cap is lost it will unravel very quickly
D. All of these choices are correct

A   When you buy a new VHF or UHF hand-held transceiver, it will usually have a flexible rubber-coated antenna commonly called a "rubber duck." This antenna is inexpensive, small, lightweight and difficult to break. On the other hand, its performance is not nearly as good as a full-sized antenna, such as a mobile antenna or a telescoping whip. [*Ham Radio License Manual*, page 4-13]

T9A05   How would you change a dipole antenna to make it resonant on a higher frequency?

A. Lengthen it
B. Insert coils in series with radiating wires
C. Shorten it
D. Add capacity hats to the ends of the radiating wires

C   Remember that wavelength and frequency change in opposite directions; as one increases, the other decreases and vice versa. To make the dipole resonant on a higher frequency, shorten it. [*Ham Radio License Manual*, page 4-12]

T9A06 **What type of antennas are the quad, Yagi, and dish?**

A. Non-resonant antennas
B. Loop antennas
C. Directional antennas
D. Isotropic antennas

C All of these antennas focus the radiated energy toward one direction. The quad and Yagi work by having groups of elements work together to reinforce radiated energy in the desired direction. The dish antenna works like a flashlight, reflecting energy so that a great deal of it is focused in one direction. [*Ham Radio License Manual*, page 4-15]

T9A07 **What is a good reason not to use a "rubber duck" antenna inside your car?**

A. Signals can be significantly weaker than when you are outside of the vehicle
B. It might cause your radio to overheat
C. The SWR might decrease, decreasing the signal strength
D. All of these choices are correct

A The vehicle's metal roof and doors act like shields, trapping the radio waves inside. Some of the signal gets out through the windows (unless they're tinted by a thin metal coating), but it's much weaker than if radiated by an external antenna. [*Ham Radio License Manual*, page 4-13]

T9A08 **What is the approximate length, in inches, of a quarter-wavelength vertical antenna for 146 MHz?**

A. 112
B. 50
C. 19
D. 12

C Start with the formula for estimating the length of a half-wavelength dipole: Length (in feet) = 468 / frequency (in MHz) = 468 / 146 = 3.2 feet. Convert to inches by multiplying by 12 = 38.5. Because this is a quarter-wavelength antenna, divide the result by two to get about 19 inches. [*Ham Radio License Manual*, page 4-11]

T9A09 **What is the approximate length, in inches, of a 6 meter 1/2-wavelength wire dipole antenna?**

A. 6
B. 50
C. 112
D. 236

C Since you already know the half-wavelength (3 meters), convert meters to feet by multiplying by 3.1 = 9.3 feet. Convert to inches by multiplying by 12 = 112 inches. [*Ham Radio License Manual*, page 4-11]

T9A10 In which direction is the radiation strongest from a half-wave dipole antenna in free space?

A. Equally in all directions
B. Off the ends of the antenna
C. Broadside to the antenna
D. In the direction of the feedline

C A dipole radiates strongest broadside to the axis of the dipole and weakest off the ends. The radiation pattern for a dipole isolated in space looks like a donut as seen in the figure below. Figure T9-1 shows both two- and three-dimensional patterns. The two-dimensional pattern shows what the three-dimensional pattern would look like if cut through the axis of the dipole. [*Ham Radio License Manual*, page 4-11]

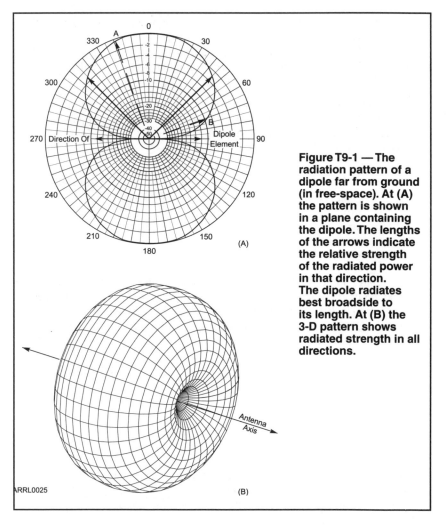

Figure T9-1 — The radiation pattern of a dipole far from ground (in free-space). At (A) the pattern is shown in a plane containing the dipole. The lengths of the arrows indicate the relative strength of the radiated power in that direction. The dipole radiates best broadside to its length. At (B) the 3-D pattern shows radiated strength in all directions.

T9A11   What is meant by the gain of an antenna?

A. The additional power that is added to the transmitter power
B. The additional power that is lost in the antenna when transmitting on a higher frequency
C. The increase in signal strength in a specified direction when compared to a reference antenna
D. The increase in impedance on receive or transmit compared to a reference antenna

C   The concentration of radio signals in a specific direction is called *gain*. (Antenna gain should not be confused with the gain of a transistor.) Antenna gain is a measure of how much signal strength is increased in a specified direction when compared to the signal from a reference antenna in the same direction. Gain aids communication in the preferred direction by increasing transmitted and received signal strengths. [*Ham Radio License Manual*, page 4-6]

T9B   Feedlines; types, losses vs. frequency, SWR concepts, matching weather protection, connectors

T9B01   Why is it important to have a low SWR in an antenna system that uses coaxial cable feedline?

A. To reduce television interference
B. To allow the efficient transfer of power and reduce losses
C. To prolong antenna life
D. All of these choices are correct

B   Power reflected from a mismatched antenna bounces back and forth along the feed line. A bit more is transferred to the antenna each time but with each pass through the feed line, some power is lost as heat. Low SWR reduces losses in the feed line because less power is reflected from the antenna. As SWR increases, more power is reflected and more power is lost. Since coaxial feed line has higher losses than open-wire line, low SWR is more important if coaxial feed line is used. [*Ham Radio License Manual*, page 4-10]

T9B02   What is the impedance of the most commonly used coaxial cable in typical amateur radio installations?

A. 8 ohms
B. 50 ohms
C. 600 ohms
D. 12 ohms

B   Most coaxial cable used in Amateur Radio has a characteristic impedance (Z_0) of 50 ohms. Coaxial cables used for video and cable television have a Z_0 of 75 ohms. Open-wire or twin-lead feed lines have a Z_0 of 300 to 450 ohms. [*Ham Radio License Manual*, page 4-9]

T9B03   Why is coaxial cable used more often than any other feedline for amateur radio antenna systems?

A. It is easy to use and requires few special installation considerations
B. It has less loss than any other type of feedline
C. It can handle more power than any other type of feedline
D. It is less expensive than any other types of feedline

A   Coaxial cable or coax is easy to work with because it is a single cable. All of the energy being conducted by the feed line is completely contained within the cable, so the cable can be run alongside or even within metallic trays, conduits and towers. It can be coiled up and placed next to other cables without effect. While it may have more loss than open-wire line and weigh a bit more per foot of length, coaxial cable's ease of use make it a practical choice for most installations. [*Ham Radio License Manual*, page 4-8]

T9B04   What does an antenna tuner do?

A. It matches the antenna system impedance to the transceiver's output impedance
B. It helps a receiver automatically tune in weak stations
C. It allows an antenna to be used on both transmit and receive
D. It automatically selects the proper antenna for the frequency band being used

A   If the SWR at the end of the feed line is too high for the radio to operate properly, devices called *impedance matchers* or *transmatches* or *antenna tuners* are connected at the output of the transmitter as shown in Figure T9-2. An antenna tuner is adjusted until the SWR measured at the transmitter output is acceptably close to 1:1. This means the antenna system's impedance has been matched to that of the transmitter output. An antenna tuner doesn't really tune the antenna, it just adjusts the impedance at the input to the antenna system. Think of the antenna tuner as an electrical gearbox that lets the engine (the transmitter) run at the speed it likes no matter how fast the tires are turning (feed point impedance). [*Ham Radio License Manual*, page 4-18]

Figure T9-2 — An antenna tuner acts like an electrical version of a mechanical gearbox. By adjusting the tuner's controls, the impedance present at the end of the feed line can be converted to the impedance that best suits the transceiver's output circuits, usually 50 ohms.

T9B05 What generally happens as the frequency of a signal passing through coaxial cable is increased?

A. The apparent SWR increases
B. The reflected power increases
C. The characteristic impedance increases
D. The loss increases

D Feed lines used at radio frequencies use special materials and construction methods to minimize power being dissipated as heat by feed line loss and to avoid signals leaking in or out. Feed line loss increases with frequency for all types of feed lines. [*Ham Radio License Manual*, page 4-8]

T9B06 Which of the following connectors is most suitable for frequencies above 400 MHz?

A. A UHF (PL-259/SO-239) connector
B. A Type N connector
C. An RS-213 connector
D. A DB-23 connector

B Which connector to use depends on the frequency of the signals being used. The UHF series of connectors are the most widely-used for HF equipment. (UHF does not stand for "ultra-high frequency" in this case.) Above 400 MHz, the Type N connectors are used. You'll find both UHF and N connectors on 6, 2 and 1.25 meter equipment. [*Ham Radio License Manual*, page 4-16]

T9B07 Which of the following is true of PL-259 type coax connectors?

A. They are good for UHF frequencies
B. They are water tight
C. They are commonly used at HF frequencies
D. They are a bayonet type connector

C See question T9B06. [*Ham Radio License Manual*, page 4-16]

T9B08 Why should coax connectors exposed to the weather be sealed against water intrusion?

A. To prevent an increase in feedline loss
B. To prevent interference to telephones
C. To keep the jacket from becoming loose
D. All of these choices are correct

A Coax connectors exposed to the weather must be carefully waterproofed because water in coaxial cable degrades the effectiveness of the braided shield and dramatically increases losses. (See question T7C09.) [*Ham Radio License Manual*, page 4-17]

T9B09   What might cause erratic changes in SWR readings?

A. The transmitter is being modulated
B. A loose connection in an antenna or a feedline
C. The transmitter is being over-modulated
D. Interference from other stations is distorting your signal

B   Remember that SWR is caused by a mismatch between feed line impedance and load or antenna impedance. If there is a loose connection where the load is attached (at the antenna feed point), then the erratic connection acts like an erratically changing load impedance. The result is an erratic change in SWR. [*Ham Radio License Manual*, page 4-10]

T9B10   What electrical difference exists between the smaller RG-58 and larger RG-8 coaxial cables?

A. There is no significant difference between the two types
B. RG-58 cable has less loss at a given frequency
C. RG-8 cable has less loss at a given frequency
D. RG-58 cable can handle higher power levels

C   An important characteristic of coax is feed line loss. Loss is specified in dB per 100 feet of cable at a certain frequency. The following table gives cable loss at 30 MHz (close to the 10 meter band) and at 150 MHz (close to the 2 meter band). In general, if made from the same materials a larger diameter cable such as RG-8 will have less loss than a small cable such as RG-58. [*Ham Radio License Manual*, page 4-16]

Cable Type	Impedance Ω	Loss per 100 feet (in dB) at 30 MHz	Loss per 100 feet (in dB) at 150 MHz
RG-8	50	1.1	2.5
RG-58	50	2.5	5.6

T9B11   Which of the following types of feedline has the lowest loss at VHF and UHF?

A. 50-ohm flexible coax
B. Multi-conductor unbalanced cable
C. Air-insulated hard line
D. 75-ohm flexible coax

C   A special type of air-insulated coaxial feed line is called *hard line* because its shield is made from a semi-flexible solid tube of aluminum or copper. This limits the amount of bending the cable can do, but hard line has the lowest loss of any type of coaxial feed line. [*Ham Radio License Manual*, page 4-8]

Electrical and RF Safety

SUBELEMENT T0 — AC power circuits, antenna installation, RF hazards [3 Exam Questions — 3 Groups]

T0A AC power circuits; hazardous voltages, fuses and circuit breakers, grounding, lightning protection, battery safety, electrical code compliance

T0A01 Which is a commonly accepted value for the lowest voltage that can cause a dangerous electric shock?

A. 12 volts
B. 30 volts
C. 120 volts
D. 300 volts

B Low-voltage power supplies may seem safe, but even battery-powered equipment should be treated with care. The minimum voltage that is considered a shock hazard to humans is 30 volts. You should respect even low voltages, taking appropriate steps to avoid contact. Lower-voltage systems such as storage batteries are quite capable of causing a fire hazard, so even without being a shock hazard, there are plenty of reasons to treat them with care. [*Ham Radio License Manual*, page 9-2]

T0A02 How does current flowing through the body cause a health hazard?

A. By heating tissue
B. It disrupts the electrical functions of cells
C. It causes involuntary muscle contractions
D. All of these choices are correct

D Burns caused by dc current or low-frequency ac current are a result of resistance to current in the skin, either through it to the body's interior or along it from point to point. The current creates heat and that's what results in the burn. Electrical current through the body can disrupt the electrical function of cells. Currents of more than a few mA can also cause involuntary muscle contractions, which leads to the jerking and jumping image on TV and in the movies. No joking matter, muscle spasms can cause falls and sudden large movements. The sudden pulling back of an outstretched hand or finger that comes in contact with an energized conductor is a result of arm muscles contracting. [*Ham Radio License Manual*, page 9-2]

T0A03 **What is connected to the green wire in a three-wire electrical AC plug?**

A. Neutral
B. Hot
C. Safety ground
D. The white wire

C State and national electrical-safety codes require the three-wire power cords on many 120-V tools and appliances. Power supplies and station equipment use similar connections. Two of the conductors (the "hot" and the "neutral" wires) power the device. The third conductor (the safety ground wire) connects to the metal frame of the device. The "hot" wire is usually black or red. The "neutral" wire is white. The frame/ground wire is green or sometimes bare. See Figure T0-1. [*Ham Radio License Manual*, page 9-4]

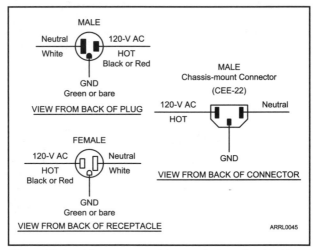

Figure T0-1 — The correct wiring technique for 120 V ac power cords and receptacles. The white wire is neutral and the green wire is the safety ground. The hot wire can be either black or red. These receptacles are shown from the back, or wiring side.

T0A04 **What is the purpose of a fuse in an electrical circuit?**

    A. To prevent power supply ripple from damaging a circuit
    B. To interrupt power in case of overload
    C. To limit current to prevent shocks
    D. All of these choices are correct

B   A fuse consists of a thin strip of metal that melts at relatively low temperatures. Current flow causes the metal to heat up and when too much current flows, the metal melts, breaking the circuit and interrupting power. [*Ham Radio License Manual*, page 3-11]

T0A05 **Why is it unwise to install a 20-ampere fuse in the place of a 5-ampere fuse?**

    A. The larger fuse would be likely to blow because it is rated for higher current
    B. The power supply ripple would greatly increase
    C. Excessive current could cause a fire
    D. All of these choices are correct

C   Never replace a fuse or circuit breaker with one rated for a larger current, since that allows more current to flow in response to a fault in the equipment. The higher current could overheat wires and cause a fire. Determine what problem caused the fuse to blow and repair it so that higher current does not make the damage worse or destroy the equipment entirely. [*Ham Radio License Manual*, page 3-11]

T0A06 **What is a good way to guard against electrical shock at your station?**

    A. Use three-wire cords and plugs for all AC powered equipment
    B. Connect all AC powered station equipment to a common safety ground
    C. Use a circuit protected by a ground-fault interrupter
    D. All of these choices are correct

D   The best way to protect against electrical shock is to make sure that each part of your station as at the same potential or voltage. That potential should, of course, be ground potential. You'll want to make sure that you have a common safety ground, meaning that all grounds should be connected together. Three-wire cords and plugs make sure the equipment enclosures are connected to the safety ground. A ground-fault interrupter detects current imbalance between the hot and neutral wires, indicating a short circuit to the safety ground, and turns off power to the circuit. [*Ham Radio License Manual*, page 9-3]

T0A07   Which of these precautions should be taken when installing devices for lightning protection in a coaxial cable feedline?

A. Include a parallel bypass switch for each protector so that it can be switched out of the circuit when running high power
B. Include a series switch in the ground line of each protector to prevent RF overload from inadvertently damaging the protector
C. Keep the ground wires from each protector separate and connected to station ground
D. Ground all of the protectors to a common plate which is in turn connected to an external ground

D   Even though amateur antennas and towers are struck no more frequently than tall trees or other nearby structures, it is wise to take some precautionary steps. This is especially true for stations in areas with frequent severe weather and lightning. Lightning protection is intended to provide fire protection for your home since most of the damage to a home resulting from a lightning strike is from fire. Starting at your antennas, all towers, masts, and antenna mounts should be grounded according to your local building codes. This is done at the base, or in the case of roof mounts, though a large-diameter wire to a ground rod. Ground connections should be as short and direct as possible — avoid sharp bends. Where cables enter the house, use lightning arrestors grounded to a common plate that is in turn connected to a nearby external ground such as a ground rod. [*Ham Radio License Manual*, page 9-4]

T0A08   What is one way to recharge a 12-volt lead-acid station battery if the commercial power is out?

A. Cool the battery in ice for several hours
B. Add acid to the battery
C. Connect the battery to a car's battery and run the engine
D. All of these choices are correct

C   Your vehicle is an excellent source of backup power. It can easily recharge a storage battery. Smaller batteries should be carefully monitored for excessive temperature when charging from a vehicle since there is no circuitry to limit charging current. Follow standard vehicle safety practices when using your car to recharge batteries. [*Ham Radio License Manual*, page 5-17]

T0A09 **What kind of hazard is presented by a conventional 12-volt storage battery?**

A. It emits ozone which can be harmful to the atmosphere
B. Shock hazard due to high voltage
C. Explosive gas can collect if not properly vented
D. All of these choices are correct

C   Storage batteries pack a lot of energy into a small volume, but with that energy comes the need to treat them carefully. The liquid acid in the battery is extremely corrosive and will eat holes in anything organic (including your clothes and skin!) During charging, hydrogen gas is given off by the battery and can be explosive if there is no ventilation to disperse it. Most of all, a short circuit in a storage battery circuit can draw hundreds of amps from the battery, melting wires and terminals and igniting insulation in a hurry! Take special care to keep tools from accidentally shorting across the battery terminals! [*Ham Radio License Manual*, page 5-17]

T0A10 **What can happen if a lead-acid storage battery is charged or discharged too quickly?**

A. The battery could overheat and give off flammable gas or explode
B. The voltage can become reversed
C. The "memory effect" will reduce the capacity of the battery
D. All of these choices are correct

A   Trying to get energy into or out of any battery too quickly can cause it to overheat. Excessive charging can cause hydrogen gas to build up faster than the battery can vent it to the outside air. Overheating is particularly dangerous with storage batteries, since they hold so much energy. Use a battery charger that is designed to work with storage batteries and keep the load on the battery to within the manufacturer's specifications. [*Ham Radio License Manual*, page 5-17]

T0A11 **Which of the following is good practice when installing ground wires on a tower for lightning protection?**

A. Put a loop in the ground connection to prevent water damage to the ground system
B. Make sure that all bends in the ground wires are clean, right angle bends
C. Ensure that connections are short and direct
D. All of these choices are correct

C   See question T0A07. [*Ham Radio License Manual*, page 9-4]

T0A12 What kind of hazard might exist in a power supply when it is turned off and disconnected?

A. Static electricity could damage the grounding system
B. Circulating currents inside the transformer might cause damage
C. The fuse might blow if you remove the cover
D. You might receive an electric shock from stored charge in large capacitors

D A capacitor is designed to store charge, so it should not be a surprise that a capacitor can remain charged even after power is removed. It is good practice to place bleeder resistors across large capacitors to slowly allow the charge to dissipate. Never assume that capacitors, particularly those in high-voltage circuits, are discharged. Measure them with a voltmeter first or use a grounding stick to discharge them. [*Ham Radio License Manual*, page 9-3]

T0A13 What safety equipment should always be included in home-built equipment that is powered from 120V AC power circuits?

A. A fuse or circuit breaker in series with the AC "hot" conductor
B. An AC voltmeter across the incoming power source
C. An inductor in series with the AC power source
D. A capacitor across the AC power source

A If you decide to run new wiring for your station as it grows, either have a licensed electrician do the wiring or inspect it. Be sure to follow the hot-black (occasionally red)/neutral-white/ground-green or bare wiring standard. Use cable and wire sufficiently rated for the expected current load. Use the proper size fuses and circuit breakers. If you build your own equipment and power it from the ac lines, be sure to always install a fuse or circuit breaker in series with the ac hot conductor. [*Ham Radio License Manual*, page 9-4]

T0B Antenna installation; tower safety, overhead power lines

T0B01 When should members of a tower work team wear a hard hat and safety glasses?

A. At all times except when climbing the tower
B. At all times except when belted firmly to the tower
C. At all times when any work is being done on the tower
D. Only when the tower exceeds 30 feet in height

C A piece of hardware or a tool will be traveling 40 mph by the time it falls 60 feet! Ouch! Whenever the crew is working on the tower, wear your protective gear — even if you are near the base of the tower. If a falling object hits the tower or a guy wire, it can bounce a long way. Try to stay out from under the crew at work at the top. And while you're at it, remember the sun block! [*Ham Radio License Manual*, page 9-13]

T0B02 What is a good precaution to observe before climbing an antenna tower?

A. Make sure that you wear a grounded wrist strap
B. Remove all tower grounding connections
C. Put on a climbing harness and safety glasses
D. All of the these choices are correct

C Make sure your climbing harness and glasses and hard hat are in good condition — then use them! Climb slowly, with a lanyard around the tower. It's not a race, so take your time. Once at the top, work slowly, thinking out each move before you make it. Have a backup plan and never work on a tower without a ground crew or someone to keep an eye on you. [*Ham Radio License Manual*, page 9-13]

T0B03 Under what circumstances is it safe to climb a tower without a helper or observer?

A. When no electrical work is being performed
B. When no mechanical work is being performed
C. When the work being done is not more than 20 feet above the ground
D. Never

D Having a ground crew is important; avoid climbing alone whenever possible because it's never safe. If you do climb alone, take along a handheld radio so that you can call for help if needed. [*Ham Radio License Manual*, page 9-13]

T0B04 Which of the following is an important safety precaution to observe when putting up an antenna tower?

A. Wear a ground strap connected to your wrist at all times
B. Insulate the base of the tower to avoid lightning strikes
C. Look for and stay clear of any overhead electrical wires
D. All of these choices are correct

C If power lines ever come into contact with your antenna, you could be electrocuted. The only safe place to install an antenna tower is in a location that is well clear of any power lines. Before you put up a tower, look for any overhead electrical wires. Make sure that the tower is installed where there is no possibility of contact between the lines and the tower if the guy wires should ever break or the tower fall. [*Ham Radio License Manual*, page 9-12]

T0B05 **What is the purpose of a gin pole?**
A. To temporarily replace guy wires
B. To be used in place of a safety harness
C. To lift tower sections or antennas
D. To provide a temporary ground

C A gin pole is a temporary mast used to lift materials such as antennas or tower sections so that you do not have to hoist things directly. This is much safer than a direct lift and supports the materials while you work on them. [*Ham Radio License Manual*, page 9-13]

T0B06 **What is the minimum safe distance from a power line to allow when installing an antenna?**
A. Half the width of your property
B. The height of the power line above ground
C. 1/2 wavelength at the operating frequency
D. So that if the antenna falls unexpectedly, no part of it can come closer than 10 feet to the power wires

D Ten feet of separation is the minimum amount for safety if the tower and antenna fall directly towards the power lines. Figure the separation from the very top of the antenna or with the antenna oriented so that it is closest to the power lines. Allow for more separation whenever possible. [*Ham Radio License Manual*, page 9-12]

T0B07 **Which of the following is an important safety rule to remember when using a crank-up tower?**
A. This type of tower must never be painted
B. This type of tower must never be grounded
C. This type of tower must never be climbed unless it is in the fully retracted position
D. All of these choices are correct

C Climbing a crank-up tower places your hands and feet between rungs and braces that can do a lot of damage if the tower sections slip or a cable breaks. Not only should you be sure the tower is fully nested, but also place a safety block such as piece of pipe or a 2-by-4 between the rungs to prevent movement. [*Ham Radio License Manual*, page 9-13]

T0B08 **What is considered to be a proper grounding method for a tower?**

A. A single four-foot ground rod, driven into the ground no more than 12 inches from the base
B. A ferrite-core RF choke connected between the tower and ground
C. Separate eight-foot long ground rods for each tower leg, bonded to the tower and each other
D. A connection between the tower base and a cold water pipe

C Don't confuse a safety ground with an RF ground! A tower's safety ground is intended to conduct any lightning energy to the earth, reducing the amount traveling along your feed lines. Grounding each leg of the tower balances lightning currents — a weak ground will encourage more energy to flow along your feed lines. It also dissipates the charge into the ground as widely as possible. [*Ham Radio License Manual*, page 9-12]

T0B09 **Why should you avoid attaching an antenna to a utility pole?**

A. The antenna will not work properly because of induced voltages
B. The utility company will charge you an extra monthly fee
C. The antenna could contact high-voltage power wires
D. All of these choices are correct

C Never attach an antenna or guy wire to a utility pole, since a mechanical failure could result in contact with high-voltage lines. [*Ham Radio License Manual*, page 9-12]

T0B10 **Which of the following is true concerning grounding conductors used for lightning protection?**

A. Only non-insulated wire must be used
B. Wires must be carefully routed with precise right-angle bends
C. Sharp bends must be avoided
D. Common grounds must be avoided

C See question T0A07. [*Ham Radio License Manual*, page 9-4]

T0B11 **Which of the following establishes grounding requirements for an amateur radio tower or antenna?**

A. FCC Part 97 Rules
B. Local electrical codes
C. FAA tower lighting regulations
D. Underwriters Laboratories' recommended practices

B See question T0A07. [*Ham Radio License Manual*, page 9-4]

TOC RF hazards; radiation exposure, proximity to antennas, recognized safe power levels, exposure to others

T0C01 What type of radiation are VHF and UHF radio signals?

A. Gamma radiation
B. Ionizing radiation
C. Alpha radiation
D. Non-ionizing radiation

D Radio and lower frequency waves are classified as *non-ionizing radiation* because the frequency is too low for them to ionize atoms, no matter how intense the power density of the wave may be. The frequency of *ionizing radiation* must be higher than that of visible light — ultraviolet, X-rays, gamma rays. Those types of radiation can separate electrons from atoms, creating *ions*. [*Ham Radio License Manual*, page 9-5]

T0C02 Which of the following frequencies has the lowest Maximum Permissible Exposure limit?

A. 3.5 MHz
B. 50 MHz
C. 440 MHz
D. 1296 MHz

B Frequencies at which the body has the highest SAR are from 30 to 1500 MHz. These are the regions on the MPE graph in Figure T0-2 where the limits for exposure are the lowest. For example, when comparing MPE for amateur bands at 3.5, 50, 440 and 1296 MHz, you can see that MPE is lowest at 50 MHz and highest at 3.5 MHz. [*Ham Radio License Manual*, page 9-6]

Maximum Permissible Exposure (MPE) Limits

Controlled Exposure (6-Minute Average)

Frequency Range (MHz)	Power Density (mW/cm^2)
0.3-3.0	(100)*
3.0-30	(900/f^2)*
30-300	1.0
300-1500	f/300
1500-100,000	5

Uncontrolled Exposure (30-Minute Average)

Frequency Range (MHz)	Magnetic Field Power Density (mW/cm^2)
0.3-1.34	(100)*
1.34-30	(180/f^2)*
30-300	0.2
300-1500	f/1500
1500-100,000	1.0

f = frequency in MHz
* = Plane-wave equivalent power density

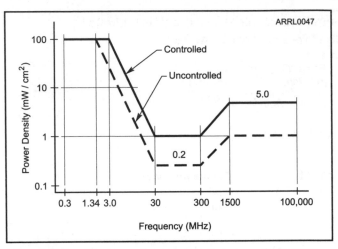

Figure T0-2 — Maximum Permissible Exposure limits vary with frequency because the body responds differently to energy at different frequencies. The controlled and uncontrolled limits refer to the environment in which people are exposed to the RF energy.

T0C03 **What is the maximum power level that an amateur radio station may use at VHF frequencies before an RF exposure evaluation is required?**

A. 1500 watts PEP transmitter output
B. 1 watt forward power
C. 50 watts PEP at the antenna
D. 50 watts PEP reflected power

C [97.13(c)(1)] — At power levels below 50 watts, the FCC has determined that there is little risk to people. Stations operating with less than 50 watts above 30 MHz are categorically excluded from having to perform a station evaluation. [*Ham Radio License Manual*, page 9-9]

T0C04 **What factors affect the RF exposure of people near an amateur station antenna?**

A. Frequency and power level of the RF field
B. Distance from the antenna to a person
C. Radiation pattern of the antenna
D. All of these choices are correct

D The human body absorbs less RF energy at some frequencies and more at others. If you decrease your transmitter output power, you decrease the RF energy radiated from your antenna. Placing antennas farther from people reduces the power density to which they are exposed. Finally, the radiation pattern of the antenna affects where RF exposure will be greatest. [*Ham Radio License Manual*, page 9-9]

T0C05   **Why do exposure limits vary with frequency?**

A. Lower frequency RF fields have more energy than higher frequency fields
B. Lower frequency RF fields do not penetrate the human body
C. Higher frequency RF fields are transient in nature
D. The human body absorbs more RF energy at some frequencies than at others

D   At frequencies near the body's natural resonant frequency, RF energy is absorbed more efficiently and maximum heating occurs. In adults, this frequency usually is about 35 MHz if the person is grounded, and about 70 MHz if the person's body is insulated from the ground. Also, body parts may be resonant as well; the adult head, for example is resonant around 400 MHz. Body size thus determines the frequency at which most RF energy is absorbed. As the frequency is increased above resonance, less RF heating generally occurs. [*Ham Radio License Manual*, page 9-6]

T0C06   **Which of the following is an acceptable method to determine that your station complies with FCC RF exposure regulations?**

A. By calculation based on FCC OET Bulletin 65
B. By calculation based on computer modeling
C. By measurement of field strength using calibrated equipment
D. All of these choices are correct

D   [97.13(c)(1)] — You may use a variety of methods to determine that your station complies with FCC RF-exposure regulations. All of the choices given above are correct and valid methods of making that determination. [*Ham Radio License Manual*, page 9-9]

T0C07   **What could happen if a person accidentally touched your antenna while you were transmitting?**

A. Touching the antenna could cause television interference
B. They might receive a painful RF burn
C. They might develop radiation poisoning
D. All of these choices are correct

B   An RF burn is caused by localized heating of the body at the point of contact with the antenna. While painful, it is rarely serious. This is a good reason why you should install your antenna where people cannot accidentally come in contact with it. [*Ham Radio License Manual*, page 9-6]

T0C08 Which of the following actions might amateur operators take to prevent exposure to RF radiation in excess of FCC-supplied limits?

A. Relocate antennas
B. Relocate the transmitter
C. Increase the duty cycle
D. All of these choices are correct

A Anything you can do to reduce power density in an area of concern will reduce RF exposure. This includes relocating your antennas farther from the area, changing the antenna's radiation pattern, and reducing power. The exposure limits also change with frequency, so changing frequency to operate on a band with a higher safe exposure limit is also acceptable. Figure T0-3 shows some options. [*Ham Radio License Manual*, page 9-10]

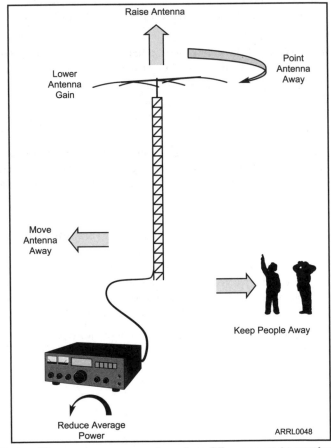

Figure T0-3 — There are many ways to reduce RF exposure to nearby people. Whatever lowers the power density in areas where people are will work. Raising the antenna will even benefit your signal strength to other stations as it lowers power density on the ground!

T0C09   How can you make sure your station stays in compliance with RF safety regulations?

A. By informing the FCC of any changes made in your station
B. By re-evaluating the station whenever an item of equipment is changed
C. By making sure your antennas have low SWR
D. All of these choices are correct

B   Whenever you make a change to something that affects power density around your antennas, you should re-evaluate the station. For example, adding an amplifier or changing to an antenna that has more gain will increase power density and you should re-evaluate. If your station is already in compliance and you take a step that decreases RF exposure, such as raising your antennas farther from areas where people are, you don't need to re-evaluate. [*Ham Radio License Manual*, page 9-9]

T0C10   Why is duty cycle one of the factors used to determine safe RF radiation exposure levels?

A. It affects the average exposure of people to radiation
B. It affects the peak exposure of people to radiation
C. It takes into account the antenna feedline loss
D. It takes into account the thermal effects of the final amplifier

A   Because the effect of RF exposure is heating, average exposure is what is important. Average exposure during any given period depends on how long the transmitter is operating — which is measured by *duty cycle*. An emission with a lower duty cycle produces less RF exposure for the same PEP output. Duty cycle is the ratio of the transmitted signal's on-air time to total operating time during the measurement period and has a maximum of 100%. (*Duty factor* is the same as duty cycle expressed as a fraction, instead of percent, such as 0.25 instead of 25%.) [*Ham Radio License Manual*, page 9-8]

T0C11   What is meant by "duty cycle" when referring to RF exposure?

A. The difference between lowest usable output and maximum rated output power of a transmitter
B. The difference between PEP and average power of an SSB signal
C. The ratio of on-air time to total operating time of a transmitted signal
D. The amount of time the operator spends transmitting

C   See question T0C10. [*Ham Radio License Manual*, page 9-8]

The ARRL Tech Q&A

A-182   Advertising

About the ARRL _____

The seed for Amateur Radio was planted in the 1890s, when Guglielmo Marconi began his experiments in wireless telegraphy. Soon he was joined by dozens, then hundreds, of others who were enthusiastic about sending and receiving messages through the air—some with a commercial interest, but others solely out of a love for this new communications medium. The United States government began licensing Amateur Radio operators in 1912.

By 1914, there were thousands of Amateur Radio operators—hams—in the United States. Hiram Percy Maxim, a leading Hartford, Connecticut inventor and industrialist, saw the need for an organization to band together this fledgling group of radio experimenters. In May 1914 he founded the American Radio Relay League (ARRL) to meet that need.

Today ARRL, with approximately 150,000 members, is the largest organization of radio amateurs in the United States. The ARRL is a not-for-profit organization that:
* promotes interest in Amateur Radio communications and experimentation
* represents US radio amateurs in legislative matters, and
* maintains fraternalism and a high standard of conduct among Amateur Radio operators.

At ARRL headquarters in the Hartford suburb of Newington, the staff helps serve the needs of members. ARRL is also International Secretariat for the International Amateur Radio Union, which is made up of similar societies in 150 countries around the world.

ARRL publishes the monthly journal *QST*, as well as newsletters and many publications covering all aspects of Amateur Radio. Its headquarters station, W1AW, transmits bulletins of interest to radio amateurs and Morse code practice sessions. The ARRL also coordinates an extensive field organization, which includes volunteers who provide technical information and other support services for radio amateurs as well as communications for public-service activities. In addition, ARRL represents US amateurs with the Federal Communications Commission and other government agencies in the US and abroad.

Membership in ARRL means much more than receiving *QST* each month. In addition to the services already described, ARRL offers membership services on a personal level, such as the ARRL Volunteer Examiner Coordinator Program and a QSL bureau.

Full ARRL membership (available only to licensed radio amateurs) gives you a voice in how the affairs of the organization are governed. ARRL policy is set by a Board of Directors (one from each of 15 Divisions) elected by the full members they represent. The day-to-day operation of ARRL HQ is managed by a Chief Executive Officer and his staff.

No matter what aspect of Amateur Radio attracts you, ARRL membership is relevant and important. There would be no Amateur Radio as we know it today were it not for the ARRL. We would be happy to welcome you as a member! (An Amateur Radio license is not required for Associate Membership.) For more information about ARRL and answers to any questions you may have about Amateur Radio, write or call:

ARRL—The national association for Amateur Radio
225 Main Street
Newington CT 06111-1494

Voice: 860-594-0200
Fax: 860-594-0259

E-mail: **hq@arrl.org**
Internet: **www.arrl.org/**

Prospective new amateurs call (toll-free):
800-32-NEW HAM (800-326-3942)
You can also contact us via e-mail at **newham@arrl.org**
or check out **ARRLWeb** at **www.arrl.org/**

FEEDBACK

We're interested in hearing your comments on this book and what you'd like to see in future editions. Please email comments to us at **pubsfdbk@arrl.org**, including your name, call sign, email address and the title, edition and printing of this book. Or you can copy this form, fill it out, and mail to ARRL, 225 Main St, Newington, CT 06111-1494.

Please check the box that best answers these questions:
How well did this book prepare you for your exam?
    ☐ Very Well   ☐ Fairly Well   ☐ Not Very Well
Did you pass?   ☐ Yes   ☐ No
Where did you purchase this book?
    ☐ From ARRL directly   ☐ From an ARRL dealer

Is there a dealer who carries ARRL publications within:
    ☐ 5 miles   ☐ 15 miles   ☐ 30 miles of your location? ☐ Not sure.

If licensed, what is your license class? _____

Name _____   ARRL member? ☐ Yes ☐ No
_____   Call Sign _____

Address _____

City, State/Province, ZIP/Postal Code _____

Daytime Phone ( ) _____ Age _____ E-mail _____

If licensed, how long? _____

Other hobbies _____

Occupation _____

For ARRL use only	Tech Q&A
Edition	·5 6 7 8 9 10 11
Printing	1 2 3 4 5 6 7 8 9 10 11